MODELLING MORPHOLOGICAL RESPONSE OF LARGE TIDAL INLET SYSTEMS TO SEA LEVEL RISE

Modelling Morphological Response of Large Tidal Inlet Systems to Sea Level Rise

DISSERTATION

Submitted in fulfillment of the requirements of
the Board for Doctorates of Delft University of Technology
and of
the Academic Board of the UNESCO-IHE Institute for Water Education
for the Degree of DOCTOR
to be defended in public
on Monday, 12 December 2011 at 15:00 hours
in Delft, the Netherlands

by

Pushpa Kumara, DISSANAYAKE
born in Onegama, Sri Lanka

Bachelor of Engineering, University of Peradeniya, Sri Lanka
Master of Science, UNESCO-IHE Institute for Water Education, Delft, the Netherlands

This dissertation has been approved by the supervisor:
Prof. dr. ir. J.A. Roelvink

co supervisor: Dr. R.W.M.R.J.B. Ranasinghe

Composition of Doctoral Committee:

Chairman	Rector Magnificus, TU Delft, the Netherlands
Prof. dr. A. Szöllösi-Nagy	Vice-Chairman, Rector, UNESCO-IHE
Prof. dr. ir. J.A. Roelvink	UNESCO-IHE/TU Delft, supervisor
Dr. R.W.M.R.J.B. Ranasinghe	UNESCO-IHE / TU Delft, co supervisor
Prof. dr. ir. A.E. Mynett	UNESCO-IHE / TU Delft
Prof. dr. ir. M.J.F. Stive	TU Delft, the Netherlands
Dr. R. Galappatti	University of Peradeniya, Sri Lanka
Dr. Z.B. Wang	TU Delft, the Netherlands
Prof. dr. ir. W.S.J. Uijttewaal	TU Delft, the Netherlands, reserve

CRC Press/Balkema is an imprint of the Taylor & Francis Group, an informa business

Published by:
CRC Press/Balkema PO Box 447, 2300 AK Leiden, the Netherlands
e-mail: Pub.NL@taylorandfrancis.com
www.crcpress.com - www.taylorandfrancis.co.uk - www.ba.balkema.nl

Cover Image: from bottom to top; schematised flat tidal inlet, 50 years evolution of the flat inlet applying tidal forcing only, 100 years evolution of the flat inlet applying tidal and wave forcings, 2004 measured bathymetry of Ameland Inlet.

ISBN 978-0-415-62100-7 (Taylor & Francis Group)

To my wife and daughters

Abstract

Large tidal inlet systems, which usually contain extensive tidal flats, are rich in bio-diversity. Economic activities and local communities in these areas have rapidly increased in recent decades. The continued existence and/or growth of these environmental systems and communities are directly linked to the existence of tidal flats which are host to a plethora of diverse flora and fauna. Alarmingly, however, these tidal flats are particularly vulnerable to any rise in mean sea level. Due to the projected future sea level rise scenarios, a clear understanding of the potential impacts of relative sea level rise (RSLR) on these systems is therefore a prerequisite for the sustainable management.

The research presented in this dissertation qualitatively investigates the morphodynamic response of a large tidal inlet/basin system to future RSLR using the state-of-the-art Delft3D numerical model and the Realistic analogue modelling philosophy (i.e. initial flat bed). The adopted approach used a highly schematised model domain representing the hydrodynamic/physical characteristics of the Ameland inlet in the Dutch Wadden Sea. The Ameland inlet is a flood-dominated inlet and consists of typical large-inlet morphological patterns (i.e. eastward oriented basin channel system, westward skewed main inlet channel and ebb-tidal delta). Applying the depth averaged version (2DH) of the Delft3D model with tidal forcing only, an established morphology which shows typical characteristics of the Ameland inlet was initially developed to investigate the inlet response to future RSLR (i.e. Eustatic sea level rise and local effects such as subsidence/rebound etc).

Model simulations were undertaken applying tidal and wave boundary forcings with three IPCC projected RSLR scenarios (No RSLR, low RSLR and high RSLR). In the first set of simulations only tidal forcing was considered. In general, RSLR accelerates the existing flood-dominance of the system leading to erosion on the ebb-tidal delta and accretion in the basin. The erosion/accretion rates are positively correlated with the rate of RSLR. Under the No RSLR scenario, the tidal flats continued to develop while they eventually drowned under the high RSLR scenario, implying that the system may degenerate into a tidal lagoon. Application of the low RSLR, resulted in more or less stable evolution of the tidal flats implying that this may be the critical RSLR for the maintenance of the system. However, this is in contrast with results obtained from the semi-empirical model (ASMITA) which indicate that the tidal flats can keep up with RSLR rates up to 10.5 mm/year. Such differences indicate the uncertainty associated with both modelling approaches.

When both tidal and wave boundary forcings were taken into account, the established morphology developed was in better agreement (compared to the tide only case) with measured Ameland bathymetry. Analysis on coastline change in comparison to the no-inlet bed (i.e. undisturbed foreshore slope) showed a higher inlet effect on the eastern barrier island (~25 km) than that on the western (~15 km) as typically found at mixed-energy tidal inlets such as Ameland inlet. Results suggested that the inlet effect on coastline change, which may be felt along the entire Ameland barrier, is mainly governed by inlet sediment bypassing, cyclic evolution of the main inlet channel and wave interaction with the ebb-tidal delta.

Sand nourishment was investigated as a mitigation strategy to counteract the RSLR induced sediment demand. Applying nourishment on the ebb-tidal delta hardly satisfied this sediment demand. However, sand nourishment on the main basin channel edge resulted in comparatively stable tidal flat evolution rather than on the ebb-tidal delta. This nourishment strategy was further investigated by applying the ASMITA model. Results indicated a stronger evolution of tidal flats in contrast to the Delft3D model.

More research focusing on the quantification of the physical and socio-economic impacts of RSLR on these systems is needed to develop effective and timely adaptation strategies enabling at least the partial preservation of bio-diversity and local communities in these regions.

Contents

Acknowledgements

This dissertation is the fruit of full-time research undertaken in Delft since February 2006. Looking back on this journey, I can remind that this research would not have been possible without direct and indirect support and encouragement of many people. Therefore, I would like to take this opportunity to express my sincere appreciation to acknowledge them.

I hope, I must begin the acknowledgement from my family who spent in Sri Lanka during the course of this research. My wife, Madhuka initially encouraged me to undertake the research knowing that she has to take care everything at home alone. The hardest decision, I have ever made is leaving the family in Sri Lanka and coming to Delft. Madhu, your support is uncountable and appreciated to the level of best. I pay my great honour to my little daughters, Mahelie and Thamodie, who missed their father for a long time and being patience.

I would like to express my deep gratitude to my promoter, Prof. J.A. Roelvink (Dano), for your countless support, inspiring feed backs throughout the research, concern being away from the family and allowing me to undertake the research partly in Sri Lanka. I initially got the privilege to work with Prof. J.A. Roelvink under my MSc research in WL| Delft Hydraulics (presently Deltares). I greatly appreciate you, Assoc. Prof. R. Hassan, for giving me this opportunity. Being my copromoter, Assoc. Prof. Rosh Ranasinghe, you tuned my research into a new perspective and I greatly appreciate your supervision, feed backs and constructive advices. I really inspired and enjoyed with your sharp comments. Dr. R. Galappatti, I cordially remind your cooperation during my tenure in Lanka Hydraulic Institute, your initial motivation to undertake the research and partly supervising my work during the stay in Sri Lanka. Dr. Z.B. Wang, I appreciate your guidance on the ASMITA model, specifically on the selection of equilibrium coefficients which is the most crucial point of this model. I greatly acknowledge the Delft Cluster research project, "*Sustainable development of North Sea and Coast (DC-05.20)*" for the financial support and Deltares for providing the Delft3D model.

Dr. Mick van der Wegen, I am really grateful to your inspiring discussions, guidance and sharing your table and even your computer from time to time. You were a lecturer for me during my MSc life, but we worked closely together in the last four years as colleagues. Ali Dastgheib, I appreciate the discussions with you in both technical and non-technical matters and your words to adapt my life from Sri Lanka to Delft. It was a great privilege for me to work some period with Dano, Mick and Ali in the same office room where I got an excellent interaction of technical matters.

Marrten van Ormondt, I have extensively used your MUPPET programme for the post-processing results and your distance support is acknowledged. Jan Joost Schouten, I appreciate your warm welcome at the Schipol airport upon returning for this research and your cooperation during the stay in Delft. I am thankful to others who helped me in the course of this work, Peter Koon Tonnon, Arjen Luijendijk, Dirk Jan Walstra, Jacco Groeneweg, André van der Westhuysen, Arjen Mol, Bert Jagers, Delft3D Support Team and others whom I unintentionally forgot.

Jolanda Boots and the student affairs staff, I admire your support in logistic and practical aspects during the research. My great appreciation goes to the library staff for their contribution and distance support during the stay away from Delft, specifically Gina Kroomt for your cooperation since my MSc life in Delft. My colleagues at UNESCO-IHE and in Delft are acknowledged for their pleasant social environments.

Finally, I express my gratitude to the present employer, Forschungsstelle Küste – NLWKN (Norderney - Germany) where I completed the last phase of the thesis. Hanz D. Niemeyer, I appreciate your understanding and allowing to undertake the final preparation partly in the office. Other colleagues specifically Ralf Kaiser, are acknowledged for their fruitful discussions and friendly office environments.

Chapter 1

Introduction

1.1   Background

Tidal inlets interrupt about 12% of the world's coastline and are associated with estuaries and tidal lagoons which are collectively defined as tidal basins (Glaeser, 1978). Barrier tidal inlets are a specific category of tidal inlets and are typically found in sandy coastal systems as a result of barrier breaching by storms and transgression of coastal plains due to sea level rise (Dronkers, 2005). A series of barrier islands is found on the US Atlantic Coast (Fenster and Dolan, 1996), the New Zealand Coast (Hicks et al., 1999; Hicks and Hume, 1996) and the Dutch, German and Danish Coast: Frisian Islands (Ehlers, 1988; Sha, 1989). The Western part of the Frisian Islands (i.e. the Dutch barrier islands) consists of large tidal inlet/basin systems (Figure 1-1).

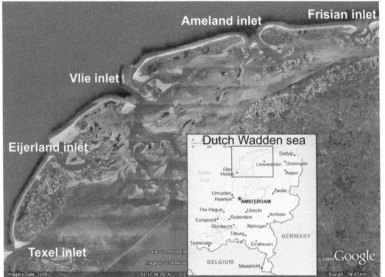

Figure 1-1 The Dutch tidal inlet/basin systems of the Wadden Sea (source: Google Earth)

Large inlet/basin systems contain extensive tidal flats which are rich in bio-diversity and associated with increase in economic activities and local communities in recent decades. The continued existence and/or growth of these environmental systems and communities are directly linked to the existence of tidal flats which are particularly vulnerable to any rise in mean sea level. In view of the projected future sea level rise scenarios, a clear understanding of the potential impacts of relative sea level rise (RSLR) on these systems is therefore a pre-requisite for their sustainable management. This research investigates the morphodynamic response of a representative tidal inlet to RSLR based on a schematised approach.

The evolution of tidal inlets is somewhat more complex than that of undisturbed coastlines. Morphological evolution of various inlet elements occurs at different spatial and temporal scales (De Vriend et al., 1991), especially where decadal/millennial evolution is concerned. The application of process-based models to investigate the long-term evolutions has now become the state-of-the-art in coastal modelling (Van der Wegen et al., 2008; Dastgheib et al., 2008). Therefore, the present research extensively uses the process-based model Delft3D to investigate RSLR induced inlet evolution. The semi-empirical model ASMITA is also employed as a basis for comparison for the process-based modelling approach.

1.2   Objective

The overarching aim of this thesis is to investigate the evolution of a tidal inlet in response to sea level rise and to evaluate possible mitigation measures for their effectiveness and potential negative impacts. The following specific research questions are formulated to achieve the main objective efficiently.

1   *Can a process-based approach be reliably applied on the decadal time scale, and does such a model predict geometric properties similar to observations?*

Process-based approaches to investigate long-term morphological changes have rapidly developed over the past decades. De Vriend et al (1993) and Latteux (1995) described reduction techniques and selection of representative conditions respectively to investigate long-term morphodynamics. More recently, Lesser et al (2004) and Roelvink (2006) have recommended modelling of long-term evolution by bed level updating at each hydrodynamic time step (i.e. *MORFAC*). This technique has been successfully applied to predict long-term evolution of tidal inlets and estuaries (Marciano et al., 2005; Van der Wegen et al., 2008; Dastgheib et al., 2008). However, numerous uncertainties, such as the sensitivity of model predictions to model domain, transport formulations, direction and asymmetry of tidal forcing and wave effect etc, when using the *MORFAC* approach have not been fully investigated to date.

2   *Does a process-based approach support the assumptions applied in semi-empirical models, such as the level of tidal flats following sea level rise with some time lag, and does this still hold under accelerated sea level rise?*

Semi-empirical models, based on empirical-equilibrium relations among inlet elements, have been used to investigate the long-term evolution of tidal inlets and to determine the critical rate of sea level rise which would result in tidal flats that are in dynamic equilibrium state (Van Goor, 2003). Further, Van Dongeren and De Vriend (1994) described the application of a semi-empirical model to predict tidal flat evolution under accelerated rates of sea level rise. However, to date, the results of these approaches have not been compared with process-based model predictions. The present study employs both the process-based model Delft3D and the semi-empirical model ASMITA to compare and contrast predictions obtained using both modelling approaches.

3   *What are the alongshore length scales of inlet influence on adjacent coastlines, and what are the dominant processes?*

Empirical formulas which are based on littoral drift and inlet hydrodynamics are used to classify tidal inlets and in turn imply shoreline effects (Bruun and Gerritsen, 1960; Hubbard, 1976; FitzGerald et al., 1978; FitzGerald, 1988; Oertel, 1988). Work and Dean (1990) used an analytical method to evaluate the inlet impact on the adjacent coastlines of the Florida Coast. A similar approach can be found in Fenster and Dolan (1996) who investigated the existing inlet effect on the US mid-Atlantic Coast. Castelle et al (2007) applied a numerical model and used aerial photographs to investigate the inlet effect on the Gold Coast, Australia from 1973 to 2005. However, there are no comprehensive studies that investigate the inlet effect on the barrier islands of the Dutch Wadden Sea.

4   *Can nourishment of the ebb-tidal delta be an effective means of feeding a tidal inlet and if so, where on the ebb-tidal delta is the most effective nourishment location?*

Sand nourishment is an increasingly adopted measure for the sustainable management of the Dutch coastal system (Stive et al., 1991; Van Duin et al., 2004; Grunnet et al., 2004 and 2005). So far, all efforts have concentrated on addressing the chronic erosion of the open coasts along the Dutch coastline. Sea level rise appears to be accelerating the sediment import into the basin. This is hypothesised to be a result of increased sediment demand by the basin due to RSLR induced increase in accommodation space. The additional supply of sand to the basin occurs at the expense of the adjacent coastlines and ebb-tidal delta, which are eroding. The efficacy of nourishing the ebb-tidal delta to mitigate these RSLR induced erosive impacts on the ebb-tidal delta itself and the adjacent coastlines and to fulfil the RSLR induced sediment demand of the tidal flats are investigated in this study.

1.3   Relevance

The Dutch Wadden Sea is characterised by a chain of barrier islands with large inlet/basin systems (Figure 1-1) which are extremely rich in biodiversity, landscape and wildlife ensuing to define as a Nature 2000 site in the EU Bird and Habitat Directives and as a UNESCO world heritage site in 2009. Therefore, these tidal basins have resulted in the entire area becoming a major tourist attraction (12 million overnight stays per year (De Jong et al., 1999)) generating significant incomes for local industries (Euro 0.7 billion per year). Furthermore, the popularity of the area has led to billions of Euros worth of development and infrastructure within the coastal zone, particularly over the last 50 years. A significant rise in the relative sea level is likely to threaten both the physical characteristics of these tidal basins and the safety of the coastal developments/infrastructures. Therefore, the present study provides an overview of possible consequences on tidal inlet morphology in response to future sea level rise scenarios and investigates potential adaptation strategies.

1.4   Thesis Structure and Approach

The structure of the thesis reflects the objectives outlined above.

Chapter 2 provides an overview of tidal inlets and the Dutch Wadden Sea area. Inlet elements and classifications are summarised. Sea level variation is described starting from the Holocene to future predictions followed by the Holocene evolution of the Dutch Wadden system. Characteristics of the study area (the Ameland inlet) are also discussed.

Chapter 3 describes the development of representative morphology to the Ameland inlet starting from the schematised flat bed. The conceptual hypotheses on inlet hydrodynamics and the process-based model (Delft3D) are summarised. Then, the selection of schematised model is discussed. Predicted hydrodynamic patterns on the flat bed are compared with conceptual hypotheses. The decadal evolution of the flat bed is investigated in terms of model/physical parameters to determine the established morphology which represents the study area.

Chapter 4 describes tidal inlet response to the future sea level rise scenarios. The sea level rise scenarios are classified considering the IPCC projections and local effects. Following the realistic analogue modelling philosophy, the established morphology is employed to investigate potential impacts of these scenarios on long-term inlet/basin evolution. Predicted bed evolutions are compared with empirical-equilibrium relations.

Chapter 5 describes inlet effect on the adjacent coastlines. Initially, the wave modules (SWAN and Xbeach) are summarised. The wave model parameters are selected based on the measured data. Application of the wave effect in long-term modelling is described in terms of a schematised wave climate. The schematised wave conditions are used to compare the performance of the wave modules. Next, the effect of wave chronology on inlet evolution is investigated. Finally, the inlet effect on the adjacent coastlines is analysed followed by a discussion of governing physical processes.

Chapter 6 describes the comparison of the process-based (Delft3D) and semi-empirical (ASMITA) model predicted inlet evolutions. Initially, the ASMITA model concept is summarised and the sensitivity of critical sea level rise to model parameters is investigated. Then, the model predicted evolutions are compared based on established and measured Ameland inlet morphologies.

Chapter 7 investigates the efficacy of sand nourishment to counter-balance the sea level rise induced sediment demand of tidal inlets. Three different strategies are investigated applying nourishment, a) *uniform nourishment*, b) *deep-area nourishment*, and c) *channel-edge nourishment*. The optimum strategy is further investigated applying the ASMITA model with the established and Ameland inlet morphologies.

Chapter 8 summarises the key results and findings of this study and identifies areas requiring further research.

Chapter 2

Tidal inlets in the Dutch Wadden Sea

2.1   What is a tidal inlet?

Elements of a tidal inlet

The primary elements of a tidal inlet are: a) flood-tidal delta in the basin (i.e. tidal flats and channels), b) inlet gorge, c) ebb-tidal delta at seaward end, and d) adjacent coastlines (Figure 2-1).

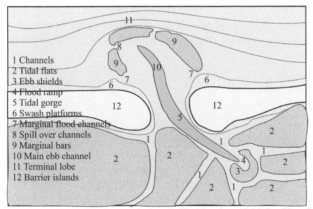

1 Channels
2 Tidal flats
3 Ebb shields
4 Flood ramp
5 Tidal gorge
6 Swash platforms
7 Marginal flood channels
8 Spill over channels
9 Marginal bars
10 Main ebb channel
11 Terminal lobe
12 Barrier islands

Figure 2-1 Typical elements of a Dutch Wadden Sea tidal inlet (Van Goor, 2001)

a.   Flood-tidal delta

The flood-tidal delta encloses the morphological features in the basin which are characterised by braided or branching channel system, inter-tidal sand and mud flats and salt marshes. Along the Dutch barrier island system, the basins are often rectangular or nearly square in shape. The morphology of the flood-tidal delta can be described in terms of channels and tidal flats. In fact, this is one system in which channels and tidal flats influence each other.

Channels

Channels are found below Mean Low Water (MLW) in the basin and often act to transport sediment and water to and from the tidal flats. The channel system in the Dutch Wadden Sea shows branching patterns. The large channels branch into smaller ones and the channel length decreases logarithmically with each bifurcation, which is related to the tidal prism and drainage area (Hibma et al., 2004). The Wadden sea channel network consists of channels that branch 3 to 4 times. The smallest length scale is about 500 m (Cleveringa and Oost, 1999) while the upper limit is defined by the geological nature (Rinaldo et al., 2001). The limiting factor of the branching pattern is thought to be the water depth that appears to halve at each bifurcation leading to less drainage area. The closer these are to the inlet the greater the depth/width of the basin channels and vice versa.

A stable relationship between the channel cross-sectional area and tidal volume in the basin under the state of morphological equilibrium was suggested by O'Brien (1931). This implies that the channel volume of a certain channel cross-section relates to the tidal volume at that specific cross-section. Such relations have been found for the Wadden Sea basins and Dutch Deltas (Renger and Partenscky, 1974; Eysink, 1990; Eysink, 1991). Accordingly, the channel volume changes if the tidal volume/prism changes due to anthropogenic or natural causes. This was evident by the field observations of the Zoutkamperlaag basin after closure of the Lauwers Sea (Eysink, 1990). This relation is compared with model predictions in the present study (see *section 4.6.3*).

Tidal flats

Tidal flats or intertidal areas that generally inundate and dry during a tidal cycle, are rich in bio-diversity (e.g. feeding grounds for birds, resting-place for seals, breeding area for fish etc). The flat volume is defined as the sand volume between MLW and Mean High Water (MHW). Tidal flats are mainly governed by tidal range, basin area, basin shape and basin orientation with respect to the dominant wind direction (Eysink, 1993). The large basins which are oriented to the wind direction, allow significant wave action around High Water (HW) because of the considerable fetch length. This likely prevents growth of extensive tidal flat areas. The crest levels of most tidal flats range between MSL and MHW-0.3 m in the Dutch Wadden Sea.

b.   *Inlet gorge*

The inlet gorge is the narrow strait between the barrier islands connecting basin and sea. Water motion through the gorge occurs as a result of different forcing types (i.e. tide, waves, wind). When the water motion is mainly governed by tide, the inlet is defined as a tidal inlet (Escoffier, 1940). Based on the hydrodynamic characteristics in the gorge, sediment can be deposited in the form of flood-tidal delta on the basin side and ebb-tidal delta on the seaside. Both forms are present in the Wadden Sea inlets.

c.   *Ebb-tidal delta*

The ebb-tidal delta accounts for the morphological features on the seaside of a tidal inlet (i.e. main ebb channel, marginal bars, terminal lobe, swash plat forms and marginal flood channels) (Figure 2-1). The ebb-tidal delta volume is defined based on the no-inlet bathymetry (Walton and Adams, 1976) of which the coastal slope is assumed to be undisturbed and similar to the adjacent barrier coasts. Therefore, the morphological boundary of the ebb-tidal delta is found where there are negligible differences between the actual and no-inlet bathymetry (Figure 2-2). The Dutch Wadden Sea inlets contain large ebb-tidal deltas which are typically oriented to the west.

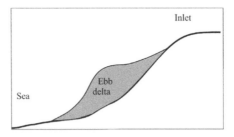

Figure 2-2 Definition of ebb-tidal delta (cross-shore profile view) relative to a no-inlet bathymetry

Tidal force

The Amplitude of the semidiurnal tide along the Dutch coast varies to a large extent due to the topography of the North Sea basin (Dronkers, 1998). The tidal regime in the North Sea is dominated by the combination of two ocean tidal waves entering from the Atlantic Ocean (Figure 2-3). These tidal waves form two amphidromic points in the North Sea rotating counter clockwise. Therefore, the tidal wave along the Dutch coast is characterised by standing and propagating waves. The first wave propagates from south to north decreasing in tidal range and merges with the second wave around Den Helder resulting in west – east propagating tidal wave which increases in range along the Wadden Sea coast (Elias, 2006; Elias et al., 2003). Accordingly, the tidal range is highest at the southern Dutch coast (3 – 4 m), the lowest at the northern central coast (1.5 – 2.0 m) with intermediate ranges along the north coast (2 – 2.5 m).

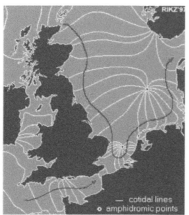

Figure 2-3 Propagation of tidal wave along the Dutch Coast (source: RIKZ)

Tidal asymmetry has a large influence on residual sediment transport in tidal inlets (Groen, 1967; Pingree and Griffith, 1979; Boon and Byrne, 1981; Aubrey and Speer, 1985; Speer and Aubrey, 1985; Dronkers, 1986; Fry and Aubrey, 1990; Van de Kreeke and Robaczewska, 1993). The asymmetry of the tidal wave is defined by the amplitude ratio (M_4/M_2) and the phase difference $(2M_2-M_4)$. The distorted tidal wave results in different durations in flood- and ebb-phases. M_4/M_2 tends to zero in the case of an undistorted tidal wave while the higher values imply higher distortion. $0^0<2M_2-M_4<180^0$ results in a flood dominant condition and $180^0<2M_2-M_4<360^0$ results in an ebb dominant condition at the inlet (Friedrichs and Aubrey, 1988). The main factors affecting tidal asymmetry in a tidal inlet are the bed topography, the reflected tidal wave from the basin and the characteristics of offshore tide. The offshore tidal asymmetry can be offset by the influence of basin geometry in long basins while dominating in short basins (Dronkers, 2005).

Classification of tidal inlets

Different classifications of tidal inlets are found based on forcing conditions (Hayes, 1979; Pugh, 1987).

A *micro-tidal range* (< 1.0 m) is commonly found in combination with relatively small
  tidal inlets. The ebb-tidal delta is marginal.
A *meso-tidal range* (low: 1.0 – 2.0 m, high: 2.0 – 3.5 m) is accompanied by smaller barrier
  islands, large inlets and large ebb-tidal deltas as in the Dutch barrier island system.
A *macro-tidal range* (low: 3.5 – 5.5 m, high: > 5.5 m) forms coastlines of tidal flats and
  flat marshes where barrier islands and ebb-tidal delta are absent (e.g. German
  Bight).

Further, Hayes (1979) classified tidal inlets into three classes (*Low*, *Medium* and *High wave energy*) based on the annual average significant wave height. The Dutch Wadden Sea has a *Medium wave energy* environment. Another classification is found based on both tidal and wave forces.

Wave dominated inlets: continuous barriers, a few tidal inlets and many washovers (e.g. Outer Banks of North Calorina, Fenster and Dolan (1996)).

Mixed energy – wave dominated inlets: many inlets and few washovers. Larger ebb-tidal deltas than the wave dominated inlets (e.g. Virginia islands, Fenster and Dolan (1996)).

Mixed energy – tide dominated inlets: many tidal inlets, larger ebb-tidal deltas and usually drumstick barrier islands (e.g. Dutch Wadden inlets, Sha (1989)).

Low – tide dominated inlets: occasionally wave built bars and transitional form (e.g. Georgia Embayment, FitzGerald (1977)).

High – tide dominated inlets: dominant tidal current ridges, salt marshes and tidal flats (e.g. Manukau inlet, North Island, New Zealand, Hicks and Hume (1996)).

A more recent classification based on ebb-tidal delta geometry, tidal range and wave force was presented by Sha and Van den Berg (1993). According to this classification, the Dutch Wadden system consists of westward oriented ebb-tidal deltas due to a west-east propagating tide, a large tidal prism and a northwesterly dominant wave.

Sediment bypassing

Sediment bypassing at an inlet depends on the wave and tide generated currents. Bruun and Gerritsen (1959) described two predominant bypassing mechanisms at tidal inlets viz. bar bypassing and tidal flow bypassing. In the case of bar bypassing, the ebb-tidal delta acts as a bridge upon which sand is carried across the inlet. Tidal flow bypassing occurs when the littoral deposits due to flood currents (landward) and waves are flushed out and spilled over by strong ebb currents (seaward). Both bypassing systems transport a large amount of sediment across the inlet via migrating sand humps or changes in channel location. The type of bypassing mechanism is defined by littoral drift to flow ratio in the inlet. High ratio (> 200 – 300) suggests bar bypassing while a low ratio (<10 – 20) implies tidal flow bypassing. Other studies have described different bypassing mechanisms at tidal inlets that are nevertheless based on these two systems (Hayes et al., 1970; Fitzgerald et al., 1976; FitzGerald et al., 1984; Michel and Howa, 1997; Kana et al., 1999; FitzGerald et al., 2000; Elias, 2006).

Inlet stability

The rate and type of bypassing are determined by the sediment supply into the inlet and the potential of flushing which depends on the tidal prism (Hubbard, 1976). Thus, the stability of a tidal inlet can be described in terms of alongshore transport and tidal prism (Bruun and Gerritsen, 1960; FitzGerald et al., 1978; Bruun, 1986). The ratio of tidal prism (Ω) to alongshore transport (M_{tot}) predicts the overall stability criterion of a tidal inlet. The Ω/M_{tot}

criterion has been validated at a great number of cases and the following ranges of the ratio have been found to be a good indicator of inlet stability.

$$r = \frac{\Omega}{M_{tot}}$$ 2-1

where, Ω in m^3 and M_{tot} in m^3/year

$r > 150$ Conditions are relatively good, little bar and good flushing

$100 < r < 150$ Conditions become less satisfactory, and offshore bar formation becomes more pronounced

$50 < r < 100$ Entrance bar may be rather large, but a channel is usually found through the bar

$20 < r < 50$ All inlets are typically 'bar-bypassers'. Waves break over the bars during storms. Inlets are stable during strong flushing.

$r < 20$ Entrances become unstable.

Later, Oertel (1988) used r together with the seaward limit of ebb jet field (*SLjf*) and littoral zone (*SLlz*) of a natural inlet to describe the state of sand bypassing system. Accordingly, four inlet systems were defined. However, none of these is found in the Wadden Sea inlets because of the higher r value in the area.

2.2   Relative sea level rise

Sea level rise affects the coastline position due to erosion/accretion depending on the local geometry and geomorphology. The world inventory of coastline changes during the last decades suggests that, of the world's sandy coastline, more than 70% are undergoing erosion, less than 10% are prograding and the remaining 20-30% are stable (Bird, 1985).

Sea level variation with respect to a fixed local bench mark is defined as *Relative Sea Level Rise* and hereon referred to RSLR which occurs as a consequence of two mechanisms. One results from global warming and climate change (i.e. thermal expansion of oceans and melting of ice caps). This is generally referred to as eustatic sea level rise (SLR). The other affects the local sea surface elevation due to vertical land movement as a result of tectonic activities or subsidence. Both mechanisms are considered in this study. RSLR will pose major challenges to long-term coastal management including erosion of coastlines/sedimentation in tidal basins.

Historical data suggest that MSL has been rising for hundreds of centuries (Louters and Gerritsen, 1994). During the last Ice Age (about 10, 000 years ago), most of the North Sea area was dry and the area of the Netherlands was free of ice. The North European ice-cap thawed (between 9000 and 8000 BP) and MSL rose by about 120 to 140 m as a result of climate becoming warmer. MSL has been gradually rising during the last few thousands years, albeit at a decreasing rate (Figure 2-4). At present, MSL appears to be increasing at a moderate rate of 14 to 17 cm/century (Rakhorst, 2000; Holgate, 2007).

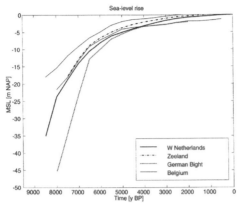

Figure 2-4 Holocene sea level rise curves for Western Netherlands (Jelgersma, 1979), Zeeland (Kiden, 1995), German Bight (Ludwig et al., 1981) and Belgium (Denys and Baeteman, 1995) after Van der Molen and Van Dijk (2000)

Recent studies by KNMI (Royal Netherlands Meteorological Institute) and IPCC (Intergovernmental Panel of Climate Change) predict that global warming and climate changes are expected to accelerate in the coming centuries as a result of increased emission of greenhouse gases (i.e. CO_2, CH_4 and NO_2). Therefore, the rate of SLR is also projected to accelerate over the next few centuries.

Several hypothetical greenhouse gas emission scenarios (35) were developed by the IPCC in their Special Report on Emission Scenarios (SRES) (IPCC 2001: Houghton et al., 2001). Applying SRES in several Atmosphere-Ocean General Circulation Models (AOGCM), the global average sea level rise from 1990-2100 has been projected (Figure 2-5). The estimated eustatic sea level rise in the next century ranges between 20 to 90 cm (IPCC 2007: Bindoff et al., 2007). The present study adopts this range of SLR projections which is sufficient to gain a qualitative understanding of SLR induced morphological changes.

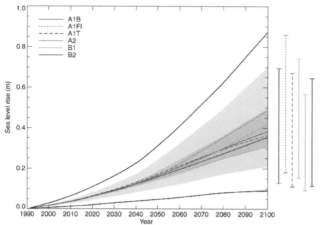

Figure 2-5 Predicted global average eustatic sea level rise from 1990 to 2100 in IPCC 2001

In addition to eustatic SLR, vertical land movement also contributes to RSLR. The most common causes of vertical land movement are compaction and tectonic activities (i.e. glacial rebound). Along the Dutch Coast, compaction plays a major role in terms of changing the deep substratum of the sea floor or consolidation of tidal flats. Additionally, land subsidence due to gas extraction also affects the MSL in the study area (Marquenie and Vlas, 2005). Presently, one gas extraction site is in operation at the east of the Ameland barrier island. This appears to have a significant influence on morphological changes in the area. The local land subsidence rate in the Ameland inlet area is expected to be 0 – 0.1 m over the next 50 years (Van der Meij and Minemma, 1999).

2.3   Historical evolution of the Wadden Sea

About 8000 years ago, the rate of SLR was a couple of meters per century, which is significantly higher than the present rate of SLR. In response to those high rates of SLR, tidal characteristics have also changed along the Dutch Coast during the Holocene period. This resulted in time varying sand transport patterns along the coast (Van der Molen and Van Dijk, 2000; Van der Molen and De Swart, 2001) leading to coastline recession (i.e. landward retreat) or progradation (i.e. seaward advance) during different periods and ultimately formed the present-day Wadden Sea barrier island system (Zagwijn, 1986; Van der Speck, 1994). The Holocene evolution of the Wadden Sea system is briefly discussed below in terms of the five main geological time periods (Zagwijn, 1986; Van der Speck, 1994).

Melting of ice caps and start of sea level rise: 100,000 to 10, 000 (C14) years ago

The North European icecap (presently Netherlands) could not be reached by sea in the last Ice Age. In fact, a thick layer of Pleistocene deposit existed on an irregular landscape. Around 15, 000 years ago, the North American ice cap started to melt resulting sea level rise of several meters per century. At this time, there was no Wadden Sea but only the Dutch dune system. Coastal evolution in response to the continuous rise in sea level resulted in the commencement of the formation of the Wadden island system.

Rapid sea level rise resulting in an oscillatory coastline: 10,000 to 7,000 (C14) years ago (9,200 – 5,800 BC)

The fragmented geological data available reveals that strong sea level rise (i.e. ~ 80 cm to few (~2 – 3) meters per century) resulted in the recession of the Dutch coastline during this period. Brackish to salt water lagoons developed in the western region of the Netherlands (Figure 2-6a). In the Wadden Sea area, estuaries with lagoons and tidal flats were formed due to flooding of river valleys and invading coastal plains. These resulted in a series of barrier islands punctuated by tidal inlets. The ever-increasing sea level and lack of alluvial sediment supply resulted in coastline recession during this period. It is cause for concern that the rates of sea level rise projected in some scenarios for the 21st century and beyond are in fact similar to that experienced during this historical period.

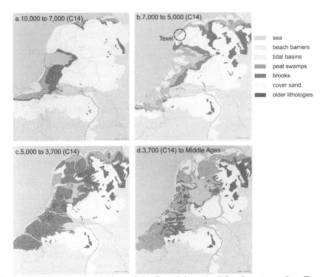

Figure 2-6 Holocene evolution of the Dutch barrier island system after Zagwijn (1986)

Formation of a fairly stable coastline: 7,000 to 5,000 (C14) years ago (5,800 to 3,780 BC).

The rate of sea level rise was about 40 to 80 cm per century during this period. Tidal flats continued shifting to landward and the Pleistocene Heights near Texel underwent severe erosion (Figure 2-6b). The Wadden Sea area during this period appears similar to the present-day system. A fairly stable coastline with tidal inlets was formed due to gradual development of barrier islands and dunes (i.e. Noord Holland and Zuid Holland provinces at present). Behind the coastline, tidal flats and salt marshes were formed covering the higher boggy areas with peat layers. During this period, the sediment supply likely satisfied demand from the additional accommodation space that resulted from the sea level rise.

Deceleration of Sea Level Rise and extension of coast: 5,000 to 3,700 (C14) years ago (3,780 to 2,100 BC).

Sea level rise decelerated to be about 20 to 40 cm per century during this period. Enough sediment was deposited on the western coast of the Netherlands due to cross-shore marine feeding (Beets and Van der Spek, 2000). This ultimately resulted in seaward migration of the coastline (Figure 2-6c). The sediment originally came from the erosion of receding capes (i.e. Zeeland delta area and Texel Heights). Later, the coast continuously eroded and receded, providing sand to the Wadden Sea basin. Then, part of the elevated flats was transformed into salt marshes or even became dry land. The Zuider Sea was not yet a sea but a freshwater lake where rivers from the south drained.

Formation of the present day Wadden environment: 3,700 (C14) years ago until the Middle Ages.

Until about 3,700 (C14) years ago, the present Wadden Sea area was similar to the western region of the Netherlands (Figure 2-6d). Thereafter, it was flooded and transformed to tidal flats. At the end of Middle Ages, shallow tidal flats developed in the elevated areas of the the Zuider Sea (i.e. the area of the present IJsselmeer). The elevated areas of the eastern Wadden Sea were flooded about 3,000 years ago leading to the eastern extension. The present-day Dutch Wadden sea system is bounded by the islands of Texel, Vlieland, Terschelling, Ameland and Frisian islands (Figure 1-1)

Human interference and natural morphological variability

In addition to sea level rise effects, human interference (i.e. diking, land reclamation, peat-cutting and damming of channels) since the Middle Ages also had a great influence on the present-day morphology of the Dutch Wadden Sea. Further, reinforcing of existing dunes to serve as dikes, construction of jetties and closing of the Zuider Sea in 1932 (Thijsse, 1972; Elias et al, 2003) resulted in major impacts on the Wadden Sea evolution.

On top of SLR induced and human induced system responses, the Wadden sea area also exhibits a large natural variability. One of the main features of this natural variability is the cyclic behaviour of main inlet channel of the Ameland inlet, whereby the system alternates between a single and double inlet channel state. This phenomenon is expected to be a result of the system's continual quest to attain the most hydraulically efficient condition at the inlet. This cyclic behaviour results in large positional changes of not only the inlet, but also the barrier islands and basin channels and tidal flats.

2.4   The Ameland inlet

Short – term characteristics

The present study investigates the evolution of the Ameland inlet which is located between the barrier islands of Terschelling (on the west) and the Ameland (on the east) (Figure 2-7). The inlet water motion and the morphology are strongly governed by offshore waves and tidal forcing. The Ameland inlet falls under the category of mixed-energy tide dominated environment (see *section 2.1*).

Tidal forcing is characterised by a semidiurnal tide with a mean tidal range of about 2.0 m propagating from West to East (~15 m/s). Therefore, the Ameland inlet experiences strong alongshore tidal currents (~0.5 – 1.0 m/s) and strong inlet currents (~1.0 m/s) (Ehlers, 1988; De Swart et al., 2004). The tidal wave shows a faster rise and slower fall which implies a flood dominant condition. The dominant wave direction is from northwest and the average significant wave height is about 1.0 m (Cheung et al., 2007). The combination of tidal and wave-induced currents leads to net easterly sediment transport between the barrier island/inlet systems.

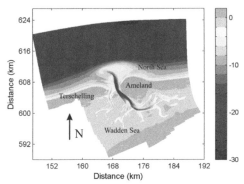

Figure 2-7 The Ameland inlet based on measured data in 2004 (De Fockert et al., 2008)

Most of the bottom material in the basin is fine sand with a mean diameter of about 0.2 mm. The basin area is about 300 km^2 while the tidal prism is approximately 480 million m^3 (Sha, 1989). Steetzel (1995) suggested that the total littoral drift from the western coast of Terschelling is in the order of 1.0 million m^3/year. Therefore, the Ameland inlet is highly stable with Ω/M_{tot} value of 480 (see *section 2.1*). This implies that tidal-flow bypassing is the dominant transport process at the inlet. A large part of the basin (~ 60%) consists of tidal flats which are submerged during high tide and partially exposed during low tide (Van Goor et al., 2003). The tidal flats restrict water exchange with adjacent basins during low tide (Vlie to the west and Frisian to the east). During high tide, the tidal flats form a tidal divide. Therefore, the basin can be considered as a relatively closed system which is convenient where numerical modelling is concerned. The main inlet channel, Borndiep, is oriented to the west at the seaward end and to the east in the basin. The inlet width is about 4 km and the maximum depth is about 27 m. The ebb-tidal delta is oriented to the west and has an area and volume of about 25 km^2 and 130 million m^3 respectively. The seaward protrusion of the ebb-tidal delta is about 6 km (Cheung et al., 2007; Wilkens, 1998).

Long- term characteristics

The Ameland inlet consists of typical morphological features viz. eastward oriented basin channel pattern, westward oriented ebb-tidal delta and main inlet channel. The inlet cyclically modulates between a one- and two-channel system with a cycle period of approximately 50 to 60 years. Israel (1998) presented a 4-state conceptual model that described this cyclic behaviour of the inlet channel configuration (Figure 2-8).

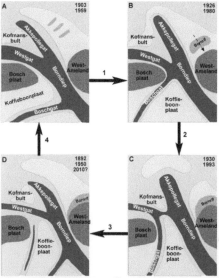

Figure 2-8 Cyclic inlet configuration hypotheses of the Ameland inlet (after Israel, 1998)

The ebb-tidal delta always consists of two channels viz. Westgat and Akkepollengat. Their size and orientation vary according to the morphological stage. The main ebb channel, Borndiep, fills and drains the eastern part of the basin and has a smooth connection with Westgat. The east-west oriented channel in the basin, Boschgat, which connects to Borndiep fills and drains the western part of the basin (Figure 2-8*A*). The dynamic behaviour of this channel results in one- and two-channel system in the inlet. The ebb-dominant flow through Westgat has a phase lag with that of Boshgat. This results in erosion of Terschelling flat and provides direct discharge from Boshgat to sea. Borndiep flat extends seawards and restricts the flow from Borndiep to Westgat leading to increased flow and sediment transport through Akkepollengat. The increased sediment supply causes the migration of the Bornrif sand bars towards the Ameland coast accomplishing the bypassing process (Figure 2-8*B*). Thereafter, the inlet gradually develops into a two-channel system (Figure 2-8*C*). At this stage, Borndiep discharges to Akkepollengat and Boshgat discharges to Westgat. Therefore, ebb-dominant flow through Westgat first decreases and eventually becomes flood-dominant. The increased sediment supply in Westgat results in rebuilding the Terschelling flat and closing north-south oriented Boshgat (Figure 2-8*D*). Ultimately, the east-west oriented Boshgat redevelops and the one-channel system is restored.

Available bathymetric data of the Ameland inlet from 1930 to 2005 were analysed to further investigate the aforementioned cyclic inlet channel behaviour. Figure 2-9 shows resulting channel patterns with contours (i.e. 0 (green), 5 (red) and 10 m (blue)). It is clear that during this time the inlet changes from a one-channel system (e.g. 1930-1980) to a two-channel system (e.g. 1990-2005). The ebb-tidal delta has two channels throughout the analysis period.

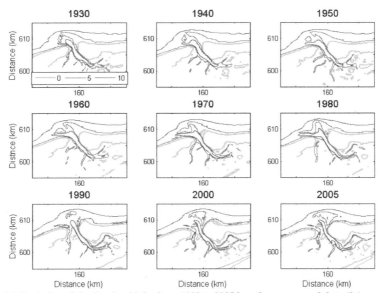

Figure 2-9 Evolution of the Ameland inlet from 1930 to 2005 based on measured data (0 (green), 5 (red) and 10 m (blue) depth contours referring to MSL)

The inlet consists of a one-channel system in 1930 and Boshgat appears to be closing at this stage (Figure 2-8*C*). Borndiep gradually develops a northward orientation at seaside leading to weaker connection with Westgat. This results in strong northward ebb currents that bypass sediment. In the mean time, the connection between Boshgat and Borndiep develops in the basin. In 1950, east-west oriented Boshgat is dominant. After 1950, Borndiep gradually develops a strong connection with Westgat. The resulting strong discharge to Westgat can be expected to cause more shoal areas on the ebb-tidal delta (i.e. 1970 and 1980). In 1990, a two-channel system is formed due to Boshgat directly discharging into sea while Borndiep maintains a northward orientation. Borndiep results in strong ebb currents on the ebb-tidal delta accomplishing the sediment bypassing process. This is evident by comparing the shoal area which has been moved towards the Ameland island on the eastern part of the ebb-tidal delta (i.e. 2000 and 2005). Unfortunately however, the data do not cover a time span that shows the complete system cycle from one-channel to two-channel to one-channel again.

Figure 2-10 shows the temporal evolution of the area-depth hypsometry. Area is defined as the wet area below the respective depth. Figure 2-10 indicates that the hypsometry hardly varies during this period. The evolution of 2 to 10 m depth is shown in Figure 2-10b. The hypsometry curves generally indicate deepening in deep areas and accretion in shallow areas during the evolution. This implies that the deep areas erode further to provide material for the accretion in shallow areas. Figure 2-10c shows the hypsometry around the tidal flat areas. In 1930, the hypsometry curve stays at the bottom and gradually moves upward during the analysis period. This is evidence that the wet area gradually decreases implying the growth of the tidal flat areas. However, it should be noted that this analysis does not consider the SLR effect which would decrease the tidal flat areas.

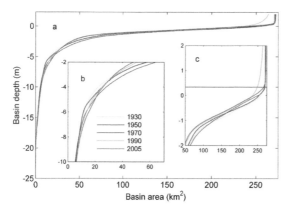

Figure 2-10 Evolution of basin-area depth hypsometry of the Ameland inlet from 1930 to 2005

Further analysis was undertaken to investigate the respective evolution of ebb-tidal delta, inlet gorge and basin. The average cross-shore profile of the ebb-tidal delta was estimated from the seaward end of the inlet. The inlet gorge evolution was analysed in terms of the middle cross-section. The basin profile was determined from the basin end of the inlet to the landward end of the basin.

The seaward extension of the ebb-tidal delta is about 6 km and the profile evolution is more or less stable (Figure 2-11a). Marginal oscillation occurs over the years due to the channel dynamic (i.e. Westgat and Akkepollengat) and sand bar formation. The profile becomes deep closer to the inlet when the channels are deep (e.g. 1970, 1980). In contrast, the shallow profiles are found at times when there is accretion on the ebb-tidal delta (e.g. 1950, 1970). The evolution of the inlet cross-sectional profile is highly dynamic compared with the ebb-tidal delta (Figure 2-11b). The main inlet channel has shifted about 1 km alongshore during this period. Two-channel systems correspond to 1990 and 2005 while a one-channel system existed at all other times during the analysis period. Figure 2-11c shows the evolution of the average cross-shore profile in the basin. The profile indicates a relatively strong variation close to the inlet due to the evolution of Borndiep and Boshgat channels and the Terschelling sand flat. Increasing of shallow areas as found with the hypsometry is apparent at the landward end of the profile. This is further evidence of sediment import into the basin which characterises a flood dominant system. Further, the SLR effect probably contributes to increased sediment import over the years. This SLR induced effect is investigated in detail during the course of the present study (see *Chapter 4*).

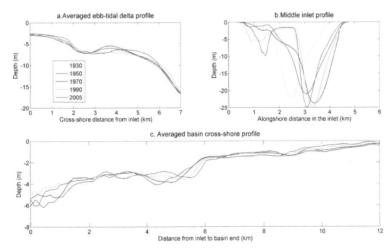

Figure 2-11 Profile evolution of the Ameland inlet; averaged cross-shore profile of ebb-tidal delta (a), middle cross-section of inlet gorge (b) and averaged cross-shore profile of basin (c)

Chapter 3

Numerical modelling of the decadal evolution of tidal inlets

Much of the material of this chapter is based on,
Dissanayake, D.M.P.K. and Roelvink, J.A., 2007. Process-based approach on tidal inlet evolution – Part 1, Proc. 5th IAHR Symposium on River, Coastal and Estuarine Morphodynamics, Twente, The Netherlands, pp. 3-9.
Dissanayake, D.M.P.K., Roelvink, J.A., Van der Wegen, M., 2009. Modelled channel pattern in a schematised tidal inlet. Coastal Engineering 59, 1069 – 1083.

3.1   Introduction

Numerous approaches describe tidal inlet evolution which results in the interactions between hydrodynamic and bed boundary processes at different spatial and temporal scales. The predictive capacity of these approaches is dependent on the model concept (i.e. data-knowledge, process-knowledge or mixture of both). Data driven models (e.g. semi-empirical/empirical) describe long-term evolution (~centuries) while process-based models are generally applied to investigate medium-term evolution (~seasonal). There appears to be a process-knowledge gap from medium- to long-term evolutions. The present analysis attempts to bridge this gap by applying a process-based model at decadal time scales and comparing predicted morphological patterns with measured data and conceptual hypotheses of inlet evolution.

Semi-empirical models predict inlet evolution over centuries (Van Goor et al., 2001). Empirical relationships show that equilibrium exists between different morphological parameters: inlet cross-sectional area and basin tidal prism (O'Brien, 1969; Jarret, 1976 and Eysink, 1990), inlet cross-sectional area and discharge (Kraus, 1998) and ebb-tidal delta volume and basin tidal prism (Walton and Adams, 1976). Wang et al (1999) described an equilibrium relation of tidal flats and inlet hydrodynamics based on Speer et al (1991) and Dronkers (1998). A number of conceptual hypotheses have been formulated to explain the morphological changes in the Dutch Wadden Sea inlets (Van Veen, 1936; Sha, 1989; Sha and Van Den Berg, 1993).

Process-based approaches to investigate long-term inlet evolutions have been rapidly developed over the past decades. De Vriend et al (1993) and Latteux (1995) described long-term morphodynamics in terms of reduction techniques and selection of representative conditions respectively. More recently Lesser et al (2004) and Roelvink (2006) have introduced the *MORFAC* technique of morphodynamic upscaling. Marciano et al (2005) used this approach to evaluate the branching channel pattern in the Wadden Sea tidal inlets.

Van der Wegen et al (2006, 2008) and Van der Wegen and Roelvink (2008) discussed long-term morphodynamics of a tidal embayment. A few process-based approaches have attempted to investigate the conceptual hypotheses (Van Leeuwen et al., 2003). However, numerous uncertainties such as the effect of model domain, transport formulations, direction and asymmetry of tidal forcing etc are yet to be investigated. The present analysis investigates these uncertainties employing the process-based model (Delft3D) which incorporates the *MORFAC* technique of morphodynamic upscaling. Inlet evolution is simulated for several decades based on a schematised approach representing the physical/hydrodynamic characteristics of the Ameland inlet in the Dutch Wadden Sea.

3.2   Conceptual hypotheses

This analysis uses three conceptual hypotheses which describe the morphological and hydrodynamic patterns of the Dutch Wadden Sea inlets.

Basin channel pattern

The first hypothesis has been formulated in the beginning of the last century (Van Veen, 1936) and relates to the back barrier basin channel pattern. This hypothesis describes why the tidal basins are characterised by an eastward oriented channel pattern (e.g. Ameland inlet). Three environmental parameters are thought to result in this morphological development; a) the tidal wave propagates into the basin from the west, b) the basin tidal prism has an eastward asymmetry with respect to the inlet, and c) the basin coast is parallel to the orientation of the back barrier islands.

Flow pattern around ebb-tidal delta

The second hypothesis describes interaction of flow pattern with the ebb-tidal delta (Sha, 1989). The ebb-tidal delta disturbs the alongshore current similar to an artificial groin, inducing a vortex on one side or the other depending on the flood/ebb situations. However, if an inlet is present, a significant vortex is only generated at the leeside during the flood tide (Figure 3-1). Sha (1989*b*) found that this is the situation in the case of Texel inlet based on the measured current pattern. The ebb-tidal delta results in a significant rotational current field only east of the inlet. This interaction enhances the westward asymmetry of the ebb-tidal delta and main inlet channel.

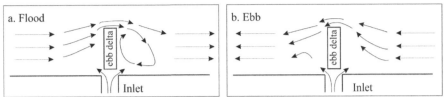

Figure 3-1 Schematic diagram of interaction alongshore tidal currents and ebb-tidal delta, a significant vortex is only generated at flood condition (after Sha, 1989)

Interaction of inlet and alongshore tidal currents

The third conceptual hypothesis describes the evolution of main inlet channel and ebb-tidal delta (Sha and Van den Berg, 1993). Interaction of shore parallel tidal currents and inlet tidal currents determines the evolution. This interaction strongly depends on local characterisation of the tidal dynamics. Types of interaction can be illustrated in four periods due to the phase difference of these currents (Figure 3-2). When the tidal range significantly differs along the coast, the alongshore water surface gradient is remarkable at low and high water levels (e.g. Vlissingen and Hoek van Holland, the Netherlands). The highest alongshore flood- and ebb-currents tend to occur at high and low water levels respectively while the inlet currents are maximum at around mid of high and low water levels. All four periods strongly contribute to the morphological evolution. In contrast, when high and low water slack occur at the same time in the inlet and adjacent offshore area (e.g. Wadden Sea Coast, the Netherlands), the shore parallel and inlet tidal currents reach their maxima at the same time around mid-tide water level (periods 2 and 4 in Figure 3-2). At flood, tidal currents are concentrated at the west of the inlet. At ebb, reversing tidal currents at sea are reinforced by the reversing tidal currents in the inlet. This again forms higher velocities west of the inlet. Further, the interaction of reversing tidal currents in the inlet and sea (period 4 in Figure 3-2) results in relatively weak and rotational tidal currents at the east of the inlet. Therefore, only periods 2 and 4 strongly contribute to morphological evolution.

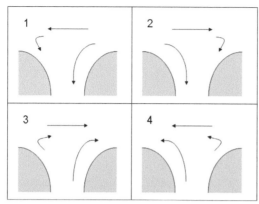

Figure 3-2 Characteristic tidal flow patterns of four different periods after Sha and Van den Berg (1993); ebb at sea and flood in inlet (1), flood at sea and flood in inlet (2), flood at sea and ebb in inlet (3), ebb at sea and ebb in inlet (4)

3.3   Process-based model, Delft3D

This study extensively uses the process-based model (Delft3D) which has been developed by Deltares (formerly WL | Delft Hydraulics) in close collaboration with Delft University of Technology. The Delft3D model consists of a number of integrated modules which together allow simulation of hydrodynamic flow (under the shallow water assumption), transport of water-borne constituents (e.g. salinity and heat), short wave generation and propagation, sediment transport and morphological changes, and ecological processes and water quality parameters.

Delft3D-FLOW module performs the hydrodynamic computations and serves as the platform in the Delft3D modelling framework. Some of the processes inside the Delft3D-FLOW module are wind shear, wave forces, tidal forces, density-driven flows and stratification due to salinity and/or temperature gradients, atmospheric pressure changes, drying and flooding of intertidal flats, sediment transport and morphological change, bed slope effects on sediment transport and bank erosion etc. Therefore, this model can be applied to a wide range of river, estuarine and coastal situations. Figure 3-3 shows different dimensions which the model can be applied in.

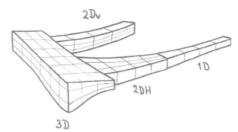

Figure 3-3 Different model dimensions of Delft3D (see FLOW User Manual)

In the present study the 2DH (depth averaged area model) version is employed because 3D processes such as vertical density stratification and curvature induced flow are not of critical importance to reach the objective of the present study. Figure 3-4 shows the structure of the Delft3D-FLOW online morphology model.

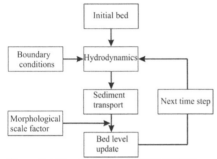

Figure 3-4 Structure of the Delft3D-FLOW module with online morphology

Hydrodynamics

In 2DH application, Delft3D-FLOW solves the unsteady shallow-water equations in two-dimensions. The system of equations consists of the continuity equation, horizontal momentum equations and a transport equation. Applications of these equations in Delft3D-FLOW are extensively described by Lesser et al (2004), and therefore are only briefly described herein.

Neglecting evaporation and precipitation, the depth-averaged continuity equation reduces to,

$$\frac{\partial \zeta}{\partial t} + \frac{\partial [h\bar{u}]}{\partial x} + \frac{\partial [h\bar{v}]}{\partial y} = 0 \qquad \text{3-1}$$

Neglecting the influence of density differences, the depth-averaged momentum equations reduce to,

$$\frac{\partial \bar{u}}{\partial t} + \bar{u}\frac{\partial \bar{u}}{\partial x} + \bar{v}\frac{\partial \bar{u}}{\partial y} + g\frac{\partial \zeta}{\partial x} + c_f \frac{\bar{u}\left|\sqrt{\bar{u}^2 + \bar{v}^2}\right|}{h} - \nu\left(\frac{\partial^2 \bar{u}}{\partial x^2} + \frac{\partial^2 \bar{u}}{\partial y^2}\right) - f_{cor}v = 0 \qquad \text{3-2}$$

$$\frac{\partial \bar{v}}{\partial t} + \bar{v}\frac{\partial \bar{v}}{\partial x} + \bar{u}\frac{\partial \bar{v}}{\partial y} + g\frac{\partial \zeta}{\partial x} + c_f \frac{\bar{v}\left|\sqrt{\bar{u}^2 + \bar{v}^2}\right|}{h} - \nu\left(\frac{\partial^2 \bar{v}}{\partial x^2} + \frac{\partial^2 \bar{v}}{\partial y^2}\right) - f_{cor}u = 0 \qquad \text{3-3}$$

$$c_f = \frac{g}{C^2} \qquad \text{3-4}$$

where,

ζ, water level (m); h, water depth (m); $\bar{u}$ and $\bar{v}$, depth averaged velocity in x and y directions (m/s); g, gravitational acceleration factor (m/s^2); c_f, friction coefficient (-); ν, eddy viscosity; C, Chèzy coefficient (m$^{1/2}$/s); f_{cor}, Coriolis parameter (1/s).

Numerical scheme

Delft3D-FLOW is a numerical model based on the finite differences. The user can select to solve the shallow water equations on a Cartesian rectangular, orthogonal curvilinear or spherical grid system. The present study uses structured Cartesian rectangular grids. The primary variables of flow (i.e. water level, velocity) are arranged on an 'Arakawa C-grid' where the water level points (pressure points) are defined at the cell center and the velocity components are perpendicular to the grid cell faces.

The unsteady shallow water equations are solved by an Alternating Direction Implicit (ADI) method which consistently estimates all parameters at each half time step (Leendertse, 1973). Delft3D-FLOW allows three options (i.e. Cyclic, Waqua and Flood) for the spatial discretisation of the horizontal advection term (Stelling, 1984). This study adopts *Cyclic* method because it is desirable for coastal environments and there is no time step limitation for advection (Stelling and Leendertse, 1991).

Sediment transport

There are several transport formulas of which one can be selected to compute sediment transport. Two distinct types of sediment transport formulas are employed in this study. One calculates the total transport as the sum of bed load transport and suspended load transport based on the depth-integrated advection-diffusion equation (Van Rijn, 1993). The other directly estimates the total transport by using the magnitude of flow velocity (Engelund and Hansen, 1967). The sensitivity of the morphological development to these formulas is investigated.

Van Rijn (1993) transport formulation

The sediment transport below and above the reference height 'a' is defined as bed load and suspended load respectively. The reference height is mainly a function of water depth and a user defined reference factor. Sediment entrainment into the water column is facilitated by imposing a reference concentration at the reference height.

Suspended sediment transport is estimated based on the advection-diffusion equation. In depth-averaged simulations, the 3D advection-diffusion equation is approximated by the depth-integrated advection-diffusion equation:

$$\frac{\partial h\bar{c}}{\partial t} + \bar{u}\frac{\partial h\bar{c}}{\partial x} + \bar{v}\frac{\partial h\bar{c}}{\partial y} - D_H\frac{\partial^2 h\bar{c}}{\partial x^2} - D_H\frac{\partial^2 h\bar{c}}{\partial y^2} = h\frac{\bar{c}_{eq} - \bar{c}}{T_s}$$

3-5

where, D_H is the horizontal dispersion coefficient (m²/s), $\bar{c}$ is the depth averaged sediment concentration (kg/m³), $\bar{c}_{eq}$ is the depth-averaged equilibrium concentration (kg/m³) as described by Van Rijn (1993) and T_s is an adaptation time-scale (s).

T_s is given by (Galappatti,1983):

$$T_s = \frac{h}{w}T_{sd}$$

3-6

where, h is the water depth, w is the sediment fall velocity and T_{sd} is an analytical function of shear velocity ($u*$) and w.

where, $u*$ is given by:

$$u'_{*,c} = (0.125 f'_c)^{0.5}\bar{u}$$

3-7

Bed load sediment transport reads:

$$|S_b| = f_{bed}\eta \times 0.5\rho_s d_{50}u'_* D_*^{-0.3}T_a$$

3-8

where, $D*$ is non-dimensional particle diameter.

$$D_* = d_{50}\left[\frac{(s-1)g}{\upsilon^2}\right]^{1/3}$$

3-9

and, T_a is the non-dimensional bed shear stress.

$$T_a = \frac{(\tau'_b - \tau_{b,cr})}{\tau_{b,cr}}$$

3-10

For *Eq.* 3-5 to 3-10, S_b, is bed load transport rate (kg/m/s); f_{bed} is a calibration factor (-); η is relative availability of sand at bottom (-); d_{50} is the mean grain diameter (m); ρ_s is density of sediment (kg/m³); f'_c is current-related friction factor (-); $\bar{u}$ is the depth average velocity (m/s); s is relative sediment density (-); υ is the horizontal eddy viscosity (m^2/s); and, $\tau_{b,cr}$ is critical bed shear stress for initiation of sediment transport (N/m^2).

d_{50} of 200 μm, ρ_s of 2650 kg/m^3 and f_{bed} of 1.0 were specified in the simulations, while a spatially varying current-related friction factor was defined as described in Van Rijn (1993).

Engelund and Hansen (1967) transport formula

This transport formula does not take into account the wave effect on the sediment motion and estimates the total load transport.

$$S = S_b + S_s = \frac{0.05\alpha U^2}{\sqrt{g}C^3\Delta^2 d_{50}}$$ 3-11

where,
α, calibration coefficient ($O(1)$); U,   magnitude of depth averaged flow velocity (m/s); Δ, relative density $(\rho_s\text{-}\rho_w)/\rho_w$; ρ_s, sediment density (kg/m^3); ρ_w, water density (kg/m^3)

Morphodynamics

Morphological acceleration factor

Coastal morphodynamic changes occur at time scales that are about 1 to 2 orders of magnitude greater than the hydrodynamic time scales (Stive et al., 1990). Therefore, in a conventional morphodynamic model, many hydrodynamic computations need to be performed to achieve significant morphological changes. Thus, morphodynamic simulations, by necessity have been very long and inefficient. However, the morphological acceleration factor (*MORFAC*) approach presented by Lesser et al (2004) and Roelvink (2006), which is used in Delft3D for bed level updates, circumnavigates this problem. In this approach, which is particularly geared at significantly improving the efficiency of morphodynamic calculations, the bed level changes calculated at each hydrodynamic time step are scaled up by multiplying erosion and deposition fluxes by a constant (*MORFAC*). Thus,

$$\Delta t_{morpho\,\log y} = MORFAC \times \Delta t_{hydrodynamic}$$ 3-12

This approach also allows accelerated bed level changes to be dynamically coupled (on-line) with the hydrodynamic computations (Figure 3-4). Therefore, long-term morphological changes can be simulated at reasonable computational cost. The main guideline to apply this approach (at present) is that the bed level changes within one tidal cycle should not exceed 10% of the local water depth.

Bed level update

The multiplied erosion and sediment fluxes after each hydrodynamic time step are used to update the bed level changes according to the sediment conservation model which solves the bed level continuity equation.

$$(1-\varepsilon)\frac{\partial z_b}{\partial t}+\frac{\partial S_x}{\partial x}+\frac{\partial S_y}{\partial y}=T_d$$

3-13

where, S_x and S_y are sediment transport components; z_b, bed level; ε, bed porosity (generally 0.4); T_d, deposition or erosion rate.

Drying and flooding

Estuaries and coastal environments consist of large, shallow and relatively flat areas. During flood tide, the entire area covers with water and the shallow areas are exposed during ebb tide. Therefore, the drying and flooding criterion (i.e. tidal flats are drying during ebb tide and becoming wet again during flood tide) is an important aspect of modelling tidal environment. Delft3D-FLOW represents this process by deactivating *dry grid points* from the flow domain during ebb tide and reactivating during flood tide after becoming *wet grid points* again. If the water depth at velocity points is less than a specified threshold depth, dry points are defined and the velocity is set to zero. If the water depth rises higher than twice the threshold depth, these points reactivated and contribute to hydrodynamic process again.

Bed slope effects

Several approaches are found to estimate the bed slope effects on the sediment transport. Present study uses the methods of Bagnold (1966) and Ikeda (1982) based on the assumption that the bed slopes have an effect only on the bed-load transport component. The influence of bed slopes is twofold. First, the magnitude of sediment transport is adjusted according to the slope along the sediment transport vector (Bagnold, 1966). Second, the sediment transport rate is adjusted based on the slope perpendicular to the sediment transport vector (Ikeda, 1982). According to Bagnold (1966), the corrected bed load transport along the longitudinal slope reads as,

$$S_b' = \alpha_s S_b$$

3-14

In which,

$$\alpha_s = 1 + \alpha_{bs}\left[\frac{\tan(\phi)}{\cos\left(\tan^{-1}\left(\dfrac{\partial z}{\partial s}\right)\right)\left(\tan(\phi) - \dfrac{\partial z}{\partial s}\right)} - 1\right]$$

3-15

S_b', corrected transport in longitudinal direction; α_s, correction factor; S_b, original bed load transport; α_{bs}, user defined longitudinal bed slope factor; ϕ, angle of repose; $\partial z / \partial s$, bed slope in the direction of bed load transport.

Subsequently, an additional bed-load transport vector is estimated perpendicular to the main bed-load transport vector following the suggestion of Ikeda (1982).

$$S_{b,n} = |S_b'|\alpha_{bn}\frac{u_{b,cr}}{|\vec{u}_b|}\cdot\frac{\partial z}{\partial n}$$

3-16

$S_{b,n}$, additional bed load transport vector; $|S_b'|$, magnitude of the transport vector adjusted for longitudinal slope only; α_{bn}, user defined transverse bed slope factor; $u_{b,cr}$, critical near-bed fluid velocity; u_b, near bed fluid velocity vector; $\partial z / \partial n$, bed slope perpendicular to the main bed-load transport vector.

Accordingly, magnitude and direction of the adjusted bed load transport can be estimated by $S_{b,n}$ and S_b'.

Bank erosion

The bank erosion process erodes dry cells depending on the sediment transport rate at adjacent wet cell sediment transport (Figure 3-5). This helps to avoid unrealistic scouring close to the dry banks. Delft3D-FLOW uses an algorithm to contribute the erosion of a wet cell to its adjacent dry cells according to the user defined percentage until the dry cells become wet cells and get activated in the hydrodynamic computation (Lesser et al., 2004; Roelvink, 2006).

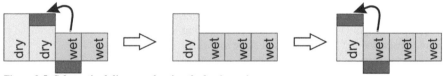

Figure 3-5: Schematised diagram showing the bank erosion process

3.4   Schematised model set-up

3.4.1   Model domain

The main objective of this analysis is to investigate the capability of process-based model to successfully simulate the decadal evolution of a large tidal inlet/basin system. As this required a number of long-term sensitivity simulations with varying physical and hydrodynamic parameters need to be undertaken several simplifying schematisations are by necessity (to reduce computational cost) adopted in this study.

The study area selected is the Ameland inlet which has a relatively closed basin in the Dutch Wadden Sea. A schematised model bathymetry is employed to represent the study area. Figure 3-6 shows the schematised model domain which consists of three rectangular areas viz. back barrier basin, inlet gorge and open sea area. It should be noted that that the schematised model domain is north-south oriented while the Ameland inlet is slightly inclined to the west (~ 10 deg.).

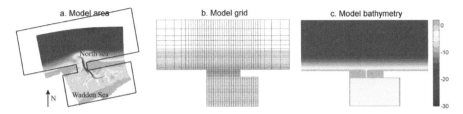

Figure 3-6 Schematisation of the Ameland inlet; Model area (a), Model grid - every 4th grid line shown (b) and Model bathymetry (c)

3.4.2   Grid set-up

There are six boundaries in the computational grid, three at the North Sea side and three at the Wadden Sea side (Figure 3-6a). The latter three boundaries were set to closed boundaries because the western and eastern boundaries lay close to the drainage divide and the southern boundary is approximately located along the basin coastline. The boundaries at the North Sea side were set to open and allow exchanging water motion with the computational domain.

Finer grid cells are applied around the inlet gorge and the open sea area consists of coarser grid cells (Figure 3-6b). Grid cell sizes in the basin are about 250 m × 100 m in x and y directions while these are about 250 m × 50 m in the inlet gorge. The maximum grid cell size is at offshore corner points (~1 km × 1 km). Additional grid cells are included at the inlet gorge to supply erodible banks (Figure 3-6c). Total number of computational grid cells is about 30,000.

3.4.3   Bathymetry

Figure 3-7 shows dimensions and a cross-section of the schematised model domain. The back barrier basin is 24 km × 13 km in x and y directions and the inlet gorge is 4 km × 3 km. The open boundaries are located sufficiently away from the area of interest resulting in 60 km × 25 km dimensions in the open sea area. The back barrier basin and inlet gorge have a flat bed of 3 m based on the averaged depth of the measured bathymetry (Figure 3-6a). The open sea area has a concave bed profile and varies from 0 m to 20 m in the first 9 km and has a constant depth of 20 m offshore. All depth values are with respect to MSL. Preliminary simulations indicated that commencing the simulations with a narrow inlet allows gradual development of the main inlet channel according to the propagation direction of the tidal wave. Therefore, the initial inlet width was set to 1 km which also agrees with the historical observations (Rijzewijk, 1981). This was implemented in the model bathymetry in terms of erodible banks. Additionally, erodible inlet banks are applied around the basin and along the North Sea coastline to investigate whether the channel pattern has a potential to extend further. The shaded area in Figure 3-7 indicates erodible banks (*dry banks*) which are elevated up to 5 m from MSL.

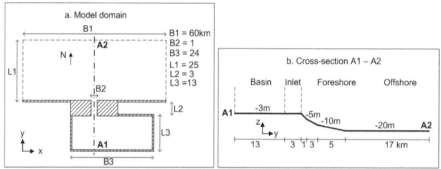

Figure 3-7 Model domain (a) and cross-section (b) of the schematised approach

3.4.4   Boundary forcing

Water motion in the Ameland inlet area is characterised by tides and waves (*section 2.1*). The present analysis adopts only tidal forcing at open boundaries. Wang et al (1995) and Ridderinkhof (1988) have suggested that major tidal constituents of this area are semi-diurnal (M_2) and quarter-diurnal (M_4). This analysis also considers the sixth-diurnal tidal component (M_6). The effect of the spring-neap cycle was not considered. Therefore, the same tidal signal repeats throughout the simulation.

The characteristics of the tidal constituents were derived by simulating a large verified 2DH model (ZUNO model) which covers the entire North Sea area (Roelvink et al., 2001). The ZUNO model was simulated together with online Fourier Analysis (FA) of the

Delft3D model. FA performs a harmonic analysis based on the user-defined harmonic components. The M_2, M_4 and M_6 constituents were extracted at offshore corner points of the schematised model domain. Subsequently, these constituents were used to define the open boundaries of the schematised Delft3D model.

The schematised model domain consists of three open boundaries and three closed boundaries (*section 3.4.1*). The velocity component perpendicular to the closed boundary was set to zero. At open boundaries, only the vertical tide was considered. This results in virtually identical net bed level changes as in the case of horizontal tide (Grunnet et al., 2005). The northern boundary was specified as a water level boundary (Table 3-1) and water level gradient boundaries (i.e. *Neumann boundary condition*) were specified at the lateral boundaries (Table 3-2 and Table 3-3). Roelvink and Walstra (2004) suggested that applying such a combination helps to minimise boundary disturbances at lateral boundaries.

Tidal component	Frequency (deg/hr)	West		East	
		Amplitude (m)	Phase (deg)	Amplitude (m)	Phase (deg)
M_2	28.9933	0.8450	20.2	0.9200	53.3
M_4	57.9866	0.0938	259.5	0.0861	304.4
M_6	86.9799	0.0625	119.5	0.0409	225.1

Table 3-1 Water level at northern boundary

The amplitude of the tidal components varies from west to east boundary (Table 3-1). This is a typical character of the propagating tidal wave along the Wadden Sea coast (Dronkers, 1986). Therefore, the Neumann conditions (i.e. water level gradient) at the lateral boundaries can not be defined as in Roelvink and Walstra (2004). The modified definition for lateral boundaries is given below.

The propagation of tidal water level along the coast reads as,

$$\eta(s,t) = \sum_{j=1}^{N} \hat{\eta}_j \cos(\omega_j t - k_j s - \varphi_j)$$ (3-17)

where; $\hat{\eta}_j$, amplitude of jth component; ω_j, angular frequency; k_j, alongshore wave number of the tidal component; s, alongshore distance; φ_j, phase relative to a fixed point in time and space.

The alongshore water level gradient is obtained by differentiating with respect to s.

$$\frac{\partial \eta}{\partial s} = \sqrt{(a^2 + b^2)}[Cos(\omega t - ks - \varphi - \alpha)]$$ (3-18)

where, $\alpha = Sin^{-1}\left(\dfrac{a}{\sqrt{(a^2 + b^2)}}\right)$ and a $= k\hat{\eta}$ and b $= \dfrac{\partial \hat{\eta}}{\partial s}$

The corrected amplitude and phase angle are $\sqrt{(a^2+b^2)}$ and ($\varphi + \alpha$) respectively.

West boundary, M$_2$

$$a = 0.845047 \times (53.3 - 20.2) \div 60000 \times \frac{\pi}{180} = 8.14137 \times 10^{-6}$$

$$b = \frac{(0.919954 - 0.845047)}{60000} = 1.24845 \times 10^{-6} \ , \ \alpha = 81.3 \,\text{deg.}$$

Length of the northern boundary is 60 km.

$Amplitude = 8.23654 \times 10^{-6}$, $Phase = 20.2 + 81.3 = 101.5 \,\text{deg.}$

Similarly, amplitudes and phases can be defined for M$_4$ and M$_6$.

Frequency	At nearshore end		At offshore end	
(deg./hr)	Amplitude (m)	Phase (deg.)	Amplitude (m)	Phase (deg.)
28.9933	8.23654e-006	101.5	8.23654e-006	101.5
57.9866	1.13307e-006	343.6	1.13307e-006	343.6
86.9799	1.23239e-006	198.9	1.23239e-006	198.9

Table 3-2 Neumann condition at west boundary

East boundary, M$_2$

$$a = 0.919954 \times (53.3 - 20.2) \div 60000 \times \frac{\pi}{180} = 8.86304 \times 10^{-6}$$

$$b = \frac{(0.919954 - 0.845047)}{60000} = 1.24845 \times 10^{-6}, \ \alpha = 82.0 \,\text{deg.}$$

$Amplitude = 8.23654 \times 10^{-6}$, $Phase = 53.5 + 82.0 = 135.5 \,\text{deg.}$

Frequency	At nearshore end		At offshore end	
(deg./hr)	Amplitude (m)	Phase (deg.)	Amplitude (m)	Phase (deg.)
28.9933	8.23654e-006	135.5	8.23654e-006	135.5
57.9866	1.13307e-006	387.9	1.13307e-006	387.9
86.9799	1.23239e-006	299.0	1.23239e-006	299.1

Table 3-3 Neumann condition at east boundary

3.4.5   Model performance

Hydrodynamic characteristics of the schematised model are compared with the ZUNO model. The results of both models are extracted at the same point which is located about 16 m depth, 7 km seaward and in-line with the inlet. The comparison uses predicted water motion and its harmonic components.

Amplitude and phase of M$_2$, M$_4$ and M$_6$ were derived applying FA on the horizontal and vertical tide. Figure 3-8 shows comparison of these parameters. Both amplitude and the phase of water level agree quite well. The ZUNO model predicts slightly higher velocity amplitude while the phases are in a good agreement. Further, it is evident that M_2 has the highest contribution on the water motion.

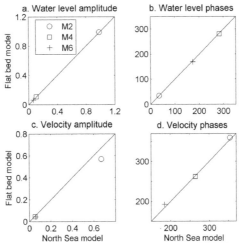

Figure 3-8 Comparison of tidal characteristics of North Sea model (ZUNO) and schematised model; Water level amplitude (a), Water level phases (b), Velocity amplitude (c) and Velocity phases (d)

Figure 3-9 shows the comparison of water level and alongshore velocity variation. The hydrodynamic characteristics of the 50-year predicted bed are also included in order to get a better insight of the schematised model (see *section 3.6*). LW is lower than that of HW in the results. This is a typical feature of tidal observations due to the *Kelvin* wave effect on the tidal wave propagation. The schematised model predicts marginal difference in water levels which is not surprising. The 50-year developed bed shows the same water level as the flat bed. This implies that the ebb-tidal delta does not significantly affect water levels.

In contrast to the water levels, the velocity prediction shows some irregularities (Figure 3-9*b*). The difference of the velocities is higher at ebb phase (negative) compare with the flood phase (positive). This is probably due to the higher bed friction on the tidal wave propagation at LW (Dronkers, 1986). It has been enhanced on the predicted 50-year bathymetry. However, during flood, the predicted bed has higher velocity similar to the ZUNO model. This could be due to the strong tidal currents being generated at the terminal lobe of the ebb-tidal delta (Sha, 1989).

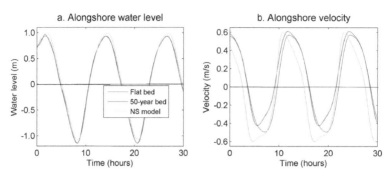

Figure 3-9 Comparison of schematised model (Flat and 50-year bed) and NS model (ZUNO); Water level (a) and Velocity (b)

Water level and velocity variation are further investigated in terms of their phase lags.
At deep water, the frictional effect on the tidal propagation is negligible and non-linear interaction plays a major role. Thus, water level and velocities are almost in phase. In contrast, the tidal wave is subjected to phase lag while propagating in the shallow water. This mainly occurs due to bed friction as a result of different bottom topography. Therefore, phase lag implies bathymetric characteristics. Figure 3-10 shows the phase lag of the water motion on the schematised and the measured bathymetries. The ZUNO model predicts a higher velocity gradient at slack water before flood and lower gradient at slack before ebb. Thus, phase lag at flood (a_1) is lower than that of ebb (b_1). Further, the inlet tidal currents and reflected tidal wave from the basin could affect the phase difference. In the schematised bathymetry, both phase lags at flood- and ebb-phases (i.e. a_2 and b_2, a_3 and b_3) appear to be similar. This implies that offshore bed topography which is very different to that in the ZUNO model, has a large influence on the phase lag compared to other effects. Further, the predicted bathymetry has the highest phase lag at flood phase. Different topography of the ebb-tidal delta probably leads to this difference.

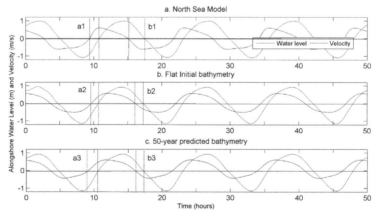

Figure 3-10 Comparison of phase lag in the North sea model (ZUNO) and the schematised model domain; Water level (red) and Velocity (blue)

The comparison of the hydrodynamic characteristics suggests that the schematised model simulated water motion patterns are very similar to those predicted by the ZUNO model at the Ameland inlet.

3.5   Initial Patterns

3.5.1   Hydrodynamic

Short-term simulation results are used to describe hydrodynamic and sediment transport characteristics. The analysis is undertaken in terms of horizontal and vertical tide, variation of M_2 velocity ellipses, mean residual flow and transport patterns.

Figure 3-11a shows the variation of alongshore and inlet tidal currents together with the offshore water level (i.e. at 17 m deep location, 7 km seaward and in-line with the inlet). The alongshore current was computed at the same point. The inlet point was selected in the middle of the inlet. Four distinct periods are observed in a tidal cycle. The numbers of the velocity periods are referred to Figure 3-2. During the long periods (2 and 4), the inlet current is inward as the alongshore current is eastward and vice versa. In contrast, during the short periods (1 and 3) the inlet current is outward when the alongshore velocity is eastward and vice versa. The water level indicates a tidal range of about 2 m. The alongshore and inlet currents reach a maximum at the mid-tide water level. Therefore, the velocity curves agree with the description of Sha and Van den Berg (1993).

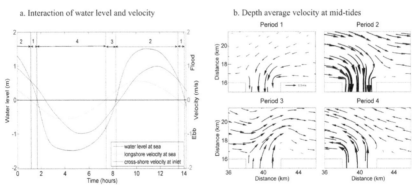

Figure 3-11 **Variation of alongshore tidal current and water level at offshore with inlet tidal current (a) and corresponding velocity pattern at mid-tide condition (b)**

In order to evaluate the interaction of tidal currents, depth averaged velocity is analysed with respect to the four velocity periods (Figure 3-11b). Period 1 shows ebb in the nearshore and flood in the inlet. This interaction develops the lowest nearshore and inlet velocities. Flood at both nearshore and in the inlet are represented by Period 2. The highest flood velocities are experienced during this period. Further, the velocities are stronger to the west of the inlet while these are weaker to the east of the inlet. Period 3 shows flood nearshore and ebb in the inlet. No significant velocities are shown in the inlet. Ebb conditions in the inlet and nearshore are represented by the last period (Period 4) which shows maximum ebb velocities which are concentrated to the west of the inlet. Relatively weaker and rotational currents are found to the east of the inlet as described by Sha (1989). Further, period 2 and 4 have longer durations and are likely result in the major morphological changes (Sha and Van den Berg, 1993).

The formation of weaker and rotational currents to the east of the inlet can be investigated via velocity ellipses. The velocity ellipse of M_2 was estimated by applying FA on each grid point. Figure 3-12 shows the computed velocity ellipses. Alongshore directed ellipses are shown away from the inlet due to strong alongshore currents. The interaction between the alongshore currents and inlet tidal currents changes the alongshore directed ellipses into cross-shore directed ellipses in front of the inlet. In the east, the alongshore directed ellipses first become more circular and then again turn into alongshore direction. This describes the rotational character of the tidal currents. The ellipses in the east are relatively small compared to the west, representing weaker currents. Therefore, the process-based model results agree with the conceptual hypothesis proposed by Sha (1989). Similar conclusions were made by Van Leeuwen et al (2003) by applying the process based hydrodynamic model HAMSOM on a schematised domain.

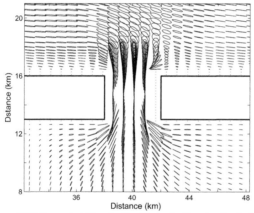

Figure 3-12 Computed M_2 velocity ellipses by Fourier Analysis

3.5.2   Residual flow and sediment transport

Figure 3-13a shows the residual flow with a maximum value in the order of 0.5 m/s. The flow pattern inside the basin is more symmetrical than that on the ebb-tidal delta. The residual flow velocities are ebb directed and asymmetrical seaward of the inlet. This current pattern is strongly influenced by the alongshore propagating tide. Therefore, a large area of landward directed velocities occurs to the west of the inlet. A rotational current field is found to the east of the inlet. This can be explained in terms of the aforementioned four velocity periods during the tidal cycle.

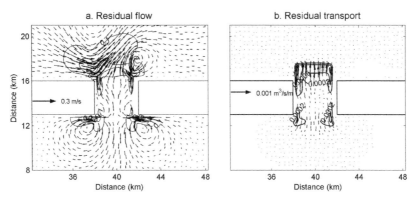

Figure 3-13 Residual flow magnitude and vectors (a), residual transport magnitude and vectors (b)

The corresponding residual sediment transport pattern was estimated for the same period as in the case of residual tide (Figure 3-13b). The transport pattern is dominated in a small area where velocities are relatively high and asymmetrical viz. at both ends of the inlet gorge. The highest transport rate is about 0.0012 m^3/s/m. The residual transport pattern is in favour of forming an ebb-tidal delta, which is a typical scenario in an inlet model with a flat basin (Dastgheib et al., 2008). This is due to the net ebb directed residual flow pattern. The locations with higher transport patterns (at both ends of the inlet gorge) are likely to develop initial channels. This type of channel formation is comparable to the model results of Van Leewen (2003).

3.6   Long-term evolutions

A number of simulations were carried out for a 50-year morphological period by varying different model settings. Initially, the sensitivity of model results to some key model parameters (i.e. *MORFAC, dry cell erosion factor* and *bed slope factors*) was analysed. The effect of *transport formulations* and *tidal asymmetry and direction* were also investigated. Finally, the bed evolution was investigated in terms of the physical set-up of the model domain (i.e. *flat basin, inlet gorge width* and *basin location relative to inlet gorge*).

3.6.1   Sensitivity of inlet evolution to model parameters

Morphological scale factor (*MORFAC*)

The present study entirely hinges upon the *MORFAC* concept to investigate long-term morphological evolutions. At present there is virtually no selection criterion for the priori determination of the highest *MORFAC* that can be used in a given situation (Ranasinghe et al., 2010). Further, this study employs a schematised model starting from a flat bed which inevitably results in large changes at the beginning of simulation, which may potentially lead to very different final outcomes based on the *MORFAC* selected (i.e. positive feedback). Therefore, a series of simulations were undertaken where the *MORFAC* was varied between 50 and 400 (i.e. 50, 100, 200 and 400). Figure 3-14 shows resulting bed evolution for each case after simulating a 50 years morphological period. In general, the higher the *MORFAC* the more diffused the developed bed patterns appear to be. All results show a westward oriented main inlet channel with a two-channel system in the inlet as found in the Ameland inlet. Further, the basin channel pattern is eastward oriented though there are some differences depending on the selected *MORFAC*. It is clearly difficult to adopt one *MORFAC* based simply on a visual comparison. Therefore, the evolutions of individual inlet elements (i.e. ebb-tidal delta, inlet and basin) were analysed further.

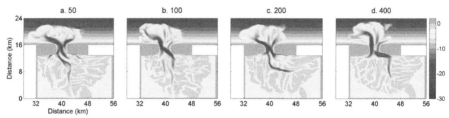

Figure 3-14 50-year bed evolutions with different *MORFAC*; 50 (a), 100 (b), 200 (c) and 400 (d)

Figure 3-15 shows volume change of the inlet elements and seaward sediment transport through the inlet gorge for the different *MORFAC*s. The ebb-tidal delta volume is defined based on the no-inlet bathymetry (Walton and Adams, 1976). The volume evolution of the inlet and basin is estimated with respect to the initial flat bed. A strong evolution of inlet elements can be seen at the beginning because of the initial flat bed with erodible banks and small inlet width. The difference of the ebb-tidal delta volume amongst the simulations is quite small (~ 5 Mm^3). In contrast, both the inlet and basin indicate relatively higher differences in volume for the different *MORFAC* applications (~ 20 Mm^3). Nevertheless, the volume evolution suggests that the integrated bed level update over the morphological period is qualitatively similar for the tested range of *MORFAC*s. Seaward transport through the inlet gorge predicts significant difference between the *MORFAC* = 400 case and the other 3 cases (Figure 3-15*d*).

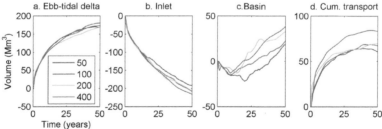

Figure 3-15 50-year evolution of inlet elements; Ebb-tidal delta (a), Inlet (b), Basin (c) and Cumulative transport in the inlet gorge (d)

Finally, the statistical method Brier Skill Score (*BSS*) (Sutherland et al., 2004; Van Rijn et al, 2003) was employed to compare predicted morphologies. Hereon, this study extensively uses *BSS* in order to compare the predicted morphological evolution with measured Ameland bathymetry.

The *BSS* is defined as follows,

$$BSS = 1 - \frac{\left\langle (z_{measured} - z_{predicted})^2 \right\rangle}{\left\langle (z_{Flat} - z_{predicted})^2 \right\rangle}$$ 3-19

where; $z_{measured}$, measured morphology of the Ameland inlet; $z_{predicted}$, modelled morphology with different *MORFAC* and z_{Flat}, the bed level of the initial flat bathymetry.

In the above definition of the *BSS*, a value of 1 indicates an excellent comparison between the measurements and model results (i.e. numerator is in the limit zero). Negative *BSS* values (i.e. numerator is large and/or denominator is small) imply large differences between the modelled and measured bathymetries. Van Rijn et al (2003) suggests the below classification for the assessment of model performance using the *BSS* (Table 3-4).

Classification	BSS
Excellent	1.0 – 0.8
Good	0.8 – 0.6
Reasonable/Fair	0.6 – 0.3
Poor	0.3 – 0.0
Bad	<0.0

Table 3-4 Classification for Brier Skill Score (Van Rijn et al., 2003)

In these simulations, the entire model domain does not develop evenly (i.e. there are areas of high and low morphological change) (Figure 3-14). To minimize the effect of small bed level differences having an undue effect on the overall *BSS,* a control area which consists only of area of interest enclosing the ebb-tidal delta, inlet and basin was isolated for *BSS* calculations. As noted previously (see *section 3.4.1*), the inlet gorge of the schematised domain is oriented north-south whereas the Ameland inlet is slightly inclined. Thus, the *BSS* was estimated for a few different orientations of the Ameland inlet (Figure 3-16).

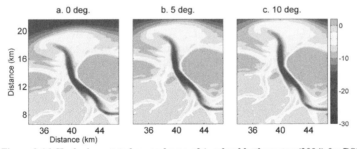

Figure 3-16 Clockwise rotated control area of Ameland bathymetry (2004) for BSS computation; 0 deg. (a), 5 deg. (b) and 10 deg. (c)

Table 3-5 shows the estimated *BSS* of *MORFACs* for different orientations of the Ameland inlet. It is apparent that the rotated control area affects the *BSS* calculations. The highest *BSS* is found with the 10 deg. rotation (Figure 3-16*c*) and a *MORFAC* of 100. However, predicted morphologies of both *MORFACs* 100 and 200 are reasonably good according to the classification of Van Rijn (2003) (Table 3-4). For this reason and the 50% decrease (compared to the *MORFAC* = 100 case) in computational time it affords, a *MORFAC* of 200 is hereon adopted to investigate the morphological evolution of the schematised bed.

Orientation of Ameland inlet (deg.)	*BSS* value of predicted morphologies of *MORFACs*			
	MORFAC=50	*MORFAC=100*	*MORFAC=200*	*MORFAC=400*
0	0.11	0.02	0.06	-0.01
5	0.33	0.29	0.31	0.18
10	0.29	0.37	0.35	0.20

Table 3-5 *BSS* value of predicted morphologies and the Ameland data with different orientation

With a *MORFAC* 200 the ebb-tidal delta volume changes from zero to about 180 Mm^3 (Figure 3-15*a*) which is, to a first order, comparable to the 2004 measured ebb-tidal delta volume of about 130 Mm^3. The discrepancy of 50 Mm^3 is likely due to the negligence of wave driven processes in the present simulations. Wave generated alongshore currents advect sediment away from the ebb-tidal delta and decrease ebb transport in the inlet. Both of these phenomena contribute to lower ebb-tidal delta volumes. The relative contribution of wave driven processes to inlet/basin evolution is discussed in *Chapter 5*.

This analysis further suggests that the differences among *MORFAC* 50, 100 and 200 predicted beds are not significant with respect to the sensitivity of the channel pattern to small disturbances (see *section 3.6.2.1*). Therefore, these three simulations are equivalent, and only the *MORFAC=400* case is significantly different than the others.

Dry cell erosion factor

Application of dry cell erosion factor is important in the evolution because the model domain consists of dry banks (see *section 3.4.3*). Sensitivity analysis was carried out by applying dry cell factors of 0, 0.5 and 1.0. In the first case, the dry banks do not contribute to the computation at all. The second case uses 50% of erosion/deposition from adjacent wet cells. In contrast, dry cells use all the amount of erosion/deposition from neighbouring wet cells in the last case. Figure 3-17 shows resulting evolutions of average cross-section of the inlet gorge. It can be seen that dry banks remain unchanged leading to an unrealistic deepening of the inlet in the first case. The other two applications result in more or less similar evolutions. However, predicted channel/shoal patterns of the basin and ebb-tidal delta are significantly different. Application of 0.5 shows a northward oriented main inlet channel, a prominent seaward extension of the ebb-tidal delta (> 8 km) and southward oriented basin channel pattern (*BSS*<0.3). In contrast, application of 1.0 results in the typical channel pattern of the Ameland inlet (*BSS*>0.3). Therefore, this study adopts a dry cell factor of 1.0.

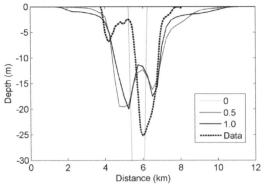

Figure 3-17 Evolution of average cross-sectional profile in the inlet with dry cell erosion factors; green (0), red (0.5) and black (1.0) measured data black-dash line

Bed slope factor

Preliminary studies with the Van Rijn transport formulas (VR) showed that the contribution of the suspended load transport to the total load transport is significantly larger (~ 90%) than the bed load transport. The Engelund and Hansen formula (EH) which estimates total load transport has been implemented in the Delft3D model as bed load transport. As described earlier, the bed slope effect is applied only to the bed load transport (see *section 3.3*). Different magnitudes of bed slope coefficients are required in these formulas to arrive at a similar downslope effect. These coefficients are uncertain and based on small scale flume experiments and can be used as calibration parameters. Sensitivity analysis with the VR formula suggested acceptable values for this study (Table 3-6).

Model	Longitudinal bed slope factor (α_{bs})	Transverse bed slope factor (α_{bn})	*BSS*
1	1	1	0.22
2	1	20	0.35
3	1	50	0.35
4	1	100	0.29

Table 3-6 Application of different longitudinal and transverse bed slope factors

Evolution of average inlet cross-sections showed that morphological changes are very sensitive to α_{bn} while being practically insensitive to α_{bs}. Cases 2 and 3 result in similar and higher *BSS* values (Table 3-6). However, Case 3 predicts a more realistic two-channel system in the inlet, while Case 2 predicts a single channel (Figure 3-18). Therefore, α_{bs} and α_{bn} of Case 3 were employed in this study. The corresponding values for the EH formula were adopted as 1 and 5 following Van der Wegen et al (2008) to ensure similar downslope effects when using either transport formulation.

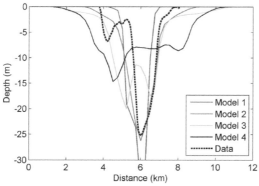

Figure 3-18 Evolution of average cross-sectional profile in the inlet with bed slope factors; $\alpha_{bs}=1$, $\alpha_{bn}=1$ (Model 1), $\alpha_{bs}=1$, $\alpha_{bn}=20$ (2), $\alpha_{bs}=1$, $\alpha_{bn}=50$ (3) and $\alpha_{bs}=1$, $\alpha_{bn}=100$ (4) and measured data

Transport formulations

Sensitivity of bed evolution to the transport formulations was investigated in terms of the VR and EH formulas. Applying the VR formula results in a more prominent bed evolution (i.e. 50 years) relative to the EH formula (Figure 3-19a,b). Predicted channel patterns with the VR formula are wider and deeper channels. Further, the ebb-tidal delta and basin show a more diffused evolution. Typical orientations of the ebb-tidal delta and the main inlet channel (i.e. westward at seaside and eastward at basin side) are found only with the VR formula. Therefore, the VR formula shows quite promising results in terms of the conceptual hypotheses of Sha (1989). However, the EH prediction indicates a symmetrical ebb-tidal delta and few channels in the inlet which are narrow and shallow. When, EH formula model was simulated up to 100 years (Figure 3-19c) the resulting channel pattern agrees better with the typical patterns predicted by the VR model (after 50 years). However, the main channels and shoals predicted by the EH model are still not as prominent as those predicted by the 50 years VR simulation.

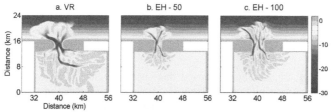

Figure 3-19 Bed evolution with transport formulas; VR – 50 years (a), EH – 50 (b) and EH – 100 (c)

Figure 3-20 shows the basin area-depth hypsometry of the predicted beds and the measured data. The maximum depth and surface area of the data are about 27 m and 270 km^2 respectively. The corresponding values of the predicted beds are about 35 m and the 300 km^2. Such relatively same differences can be explained by the highly schematised nature of the model area. In shallow water, the predicted hypsometry curves show a constant depth while the basin area increases. This is due to the fact that the basin is not yet fully developed during the morphological period. The subplot of Figure 3-20 shows that the model results agree with the data in deep water. This is more evident with the EH results (100 years) because of narrow and deep channels. The hypsometric curves indicate tidal flat areas in the basin. However, the differences between the model results are not visible. Therefore, time evolution of the flat area is further analysed.

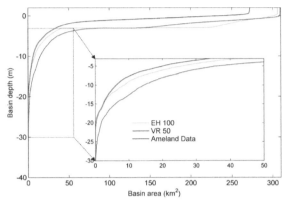

Figure 3-20 Basin area-depth hypsometry based on the predicted bed evolutions and data

The flat area in the basin indicates model sensitivity to transport formulations (Figure 3-21). After the 50-year simulation period, the corresponding values of the flat area are about 60 and 20 km^2 for VR and EH formulas respectively. This implies that the rate of sediment transport estimated by the VR formula is about three-times larger than that of the EH formula. Laboratory experiments show that the VR formula gives approximately twice the sediment transport rate as in the EH formula (Soulsby, 1997). Therefore, it appears that the VR results may be an overestimation. It should also be noted that the predicted tidal flats are not yet stable and still being developed as indicated by the high growth rate.

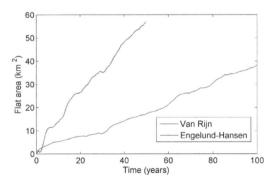

Figure 3-21 Evolution of tidal flat area with Van Rijn and Engelund-Hansen formula

The applicability of the previously developed conceptual hypotheses is investigated by analysing the hydrodynamic behavior on the predicted beds. The results show that the two large periods (refer Figure 3-11a) are still dominated by strong tidal currents to the west of the inlet which governs morphological evolution (Sha and Van den Berg, 1993). Figure 3-22 shows interaction between the alongshore tidal currents and the ebb-tidal delta at mid-flood. Both the VR (50 years) and EH (100 years) morphology produces strong currents at the terminal lobe and weaker currents to the east of the inlet due to the large seaward extension of the ebb-tidal delta. Therefore, both transport formulas are able to reproduce the conceptual hypothesis of Sha (1989) to some extent.

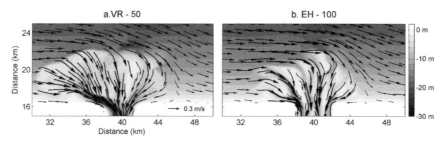

Figure 3-22 Interaction of alongshore tidal currents with ebb-tidal delta at mid-flood water level; Van Rijn formula, 50 years evolution (a) and Engelund-Hansen formula, 100 years evolution (b)

Tidal asymmetry and direction

The effect of tidal asymmetry (see *section 2.1*) was investigated by applying three scenarios (i.e. M_2 only; M_2 and M_4; M_2, M_4 and M_6) at offshore boundaries with both alongshore and cross-shore propagating tides.

Preliminary simulations predicted that scenario 2 (M_2 and M_4) and 3 (M_2, M_4 and M_6) resulted in more or less similar tidal asymmetry at the schematised inlet and as found at the Ameland inlet (i.e. $M_4/M_2 \sim O(-1)$ and $0<2M_2-M_4<180^0$). Further, the results suggested that the amplitude of the offshore applied M_4 component is one order of magnitude higher than the internally generated M_4. Figure 3-23 shows bed evolution and volume change in the inlet. Results show that the boundary conditions have a significant influence on the predicted channel pattern. However, the typical morphological features are still consistent in the results. The corresponding *BSS* values of the predicted beds are 0.26 (M_2 only), 0.17 (M_2 and M_4) and 0.35 (M_2, M_4 and M_6). The basin and ebb-tidal delta volumes are highly affected by the tidal asymmetry compared to the inlet (Figure 3-23). The basin volume decreases in the first scenario while it increases in the other two. Therefore, stronger sediment export from the inlet is evident when imposing only M_2. All models predict decreasing inlet sand volumes due to the initial configuration which most likely over-rides the effect of tidal asymmetry.

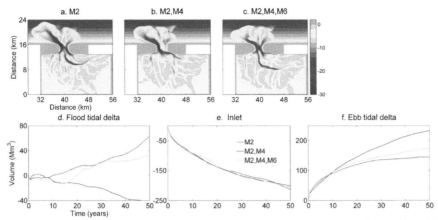

Figure 3-23 Bed evolution and volume change with different tidal boundaries; Only M₂ (a), M₂ and M₄ (b), M₂, M₄ and M₆ (c), Flood tidal delta (d), Inlet (e) and Ebb-tidal delta (f)

The 50-year bed evolution with cross-shore tidal forcing shows a much more symmetrical ebb-tidal delta (Figure 3-24). The main inlet channel branches to the west and east at the seaward end due to the symmetrical tidal forcing. However, the western branch appears to be marginally pronounced. Further, the ebb-tidal delta shows a slight asymmetry to the west. The velocity pattern is also symmetrical around the inlet, but the ebb currents are slightly more concentrated along the western channel. This can be explained by the easterly oriented basin tidal prism. Neither the hypothesis of Sha (1989) nor the hypothesis of Sha and Van den Berg (1993) are reproduced in the inlet.

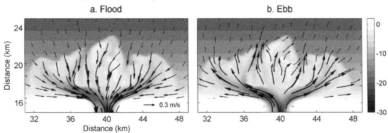

Figure 3-24 Depth average velocity at mid-tide water levels with cross-shore tidal forcing

The seaward extension of the ebb-tidal delta also shows the influence of the tidal propagation direction (Figure 3-25) on predicted bed levels. Initially, a strong extension occurs due to the flat bed initial condition. The alongshore tidal condition results in a stable value of 6 km and later extends again up to about 7 km. The alongshore propagating tide forms higher velocities at the west of the inlet due to the interaction of tidal currents (see *section 3.5.1*) which carry more sediment to the west leading to a strong initial extension. Thereafter, the main channel tends to develop to the north resulting in a stable extension. Later, the ebb-tidal delta extends again after the initial stabilization. In contrast, the cross-shore tide gradually develops the ebb-tidal delta which stabilizes at about 7 km. Therefore, both tidal directions result in similar ebb-delta extension at the end of the 50 years morphological period.

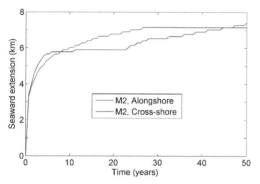

Figure 3-25 Seaward extension of the ebb-tidal delta with alongshore and cross-shore tide

The above analysis shows that it is important to consider the offshore tidal asymmetry (i.e. M_2, M_4 and M_6 resulted in the highest *BSS*) in order to investigate morphological changes. The results show that the typical channel pattern of the Ameland inlet is significantly affected by the alongshore propagating tide.

3.6.2   Sensitivity of inlet evolution to physical parameters

Three physical parameters are selected in this analysis viz. *initial flat bathymetry*, *inlet width* and *location of the basin relative to the inlet* (Table 3-7). Models are simulated with M_2, M_4 and M_6 tidal constituents and the VR transport formula together with the selected model parameters.

Parameter	Standard model	Series 1	Series 2	Series 3
Initial flat bathymetry (depth, m)	3	**Rand 1** **Rand 2** **Rand 3**	3	3
Inlet width (km)	1	1	**3.5**	1
Basin location (relative to inlet)	East	East	East	**Middle** **West**

Table 3-7 Sensitivity analysis of inlet evolution to physical parameters

3.6.2.1   Initial flat bathymetry

The effect of the initial bathymetry was analysed by applying a random perturbation of +/- 1 cm (Series 1 in Table 3-7) to the initial flat bed levels. Three randomisation schemes

(*Rand 1*, *Rand 2* and *Rand 3*) were adopted to simulate the model for a 50-year morphological period while other parameters are similar to the standard model set-up.

Figure 3-26 shows the final bed evolutions. Visual comparison suggests different channel patterns though the typical features of the Ameland inlet are consistently shown. However, it is difficult to identify the significance of the perturbation schemes on the evolution. Therefore, sub-areas are analysed separately viz. basin, inlet, ebb-tidal delta and a control area which encloses the inlet and the highly dynamic area of the basin and the ebb-tidal delta.

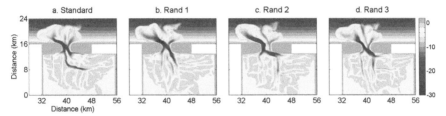

Figure 3-26 Bed evolutions with initial perturbations on the flat bed; Standard model (a), Random 1 (b), Random 2 (c) and Random 3 (d)

The temporal variations of three statistical parameters were estimated in the selected sub-areas. The first parameter is *Average depth* which is based on the grid cell area because the grid cell sizes are not uniform (z_{avg}). A decrease/increase in the average depth in the sub-area implies that sedimentation/erosion occurs. The average depth of the model domain should be constant to satisfy the requirement of mass conservation.

$$z_{avg} = \frac{\sum_{m=1}^{M}\sum_{n=1}^{N}(z(x,y).A(x,y))}{\sum_{m=1}^{M}\sum_{n=1}^{N}A(x,y)}$$

3-20

where, $z(x,y)$, grid cell depth; $A(x,y)$, grid cell area; x and y, alongshore and cross-shore coordinates; M and N, number of grid points in alongshore and cross-shore direction.

The second parameter is the *Standard deviation* of the depth (z_{sdev}) which shows the relative variation of the depth based on the average depth. The higher the variation the more variable the evolution is. This parameter will indicate the most dynamic area of the model domain.

$$z_{sdev} = \sqrt{\frac{\sum_{m=1}^{M}\sum_{n=1}^{N}(z(x,y)-z_{avg}(x,y))^{2}}{MN}}$$

3-21

The third parameter is the *Root-Mean-Square* value of the depth (z_{RMS}) with respect to the initial flat bed of the standard model. Higher/lower values imply larger/smaller evolution.

$$z_{RMS} = \sqrt{\frac{\sum_{m=1}^{M}\sum_{n=1}^{N}(z_{0,std}(x,y) - z(x,y))^2}{MN}}$$ 3-22

where, $z_{0,std}$, depth of the initial flat bed of the standard model.

Figure 3-27 shows the resulting evolution of the statistical parameters in rows and sub-areas in columns. The first column (*a*) shows that the evolution of the control area. z_{avg} is more or less constant during the evolution (see *y* scale) because the control area consists exclusively of the highly dynamic areas of the model domain. For the same reason, z_{sdev} shows a marginal increase. However, z_{RMS} shows a strong variation implying a larger evolution with respect to the flat bed. Further, a high rate of evolution is shown when the randomisation schemes are applied. Column *b*, *c* and *d* show the evolution of basin, inlet and ebb-tidal delta area respectively.

The largest evolution of z_{avg} is evident in the inlet because it has a narrow inlet gorge with erodible banks (i.e. initial level at 5 m + MSL). The basin area shows the smallest evolution (Figure 3-26). The standard model predicts slightly lower values which imply accelerated evolution with the perturbation schemes. However, this is not apparent in the ebb-tidal delta because the depth values cover a large range due to the sloping bed profile. Accelerated evolution is also seen with z_{RMS} which estimates variations with respect to the initial flat bed. The effect randomisation is largest on the evolution of the ebb-tidal delta and inlet.

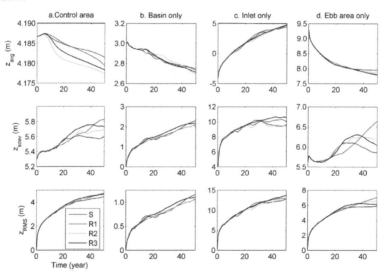

Figure 3-27 Temporal evolution of statistical parameters (z_{avg}, z_{sdev} and z_{RMS}) of the bed evolution; Control area (a), Basin (b), Inlet (c) and Ebb-tidal delta (c) in Standard model (S), Rand 1 (R1), Rand 2 (R2) and Rand 3 (R3)

The above analysis shows that small perturbations of the initial flat bathymetry result in different bed evolutions. However, all predictions are consistent with the typical features of the Ameland inlet.

3.6.2.2 Inlet width

The hydrodynamic patterns and long-term evolution of the inlet were investigated for narrow (1.0 km) and wide inlet widths (3.5 km) (Series 2 in Table 3-7). Figure 3-28 shows depth averaged velocity vectors corresponding to period 4 in Figure 3-11(mid-ebb). The maximum ebb velocity is about 2.5 m/s in the narrow inlet and it is about 1.5 m/s in the wider inlet. A strong constriction of flow is seen in the narrow inlet gorge. Both models predict that inlet and alongshore tidal currents reach maximum and minimum nearly at the same time. Weak tidal currents tend to occur to the east of the inlet and it appears to be stronger with the wider inlet. Thus, the results agree with the conceptual hypothesis of Sha (1989) irrespective of the inlet width.

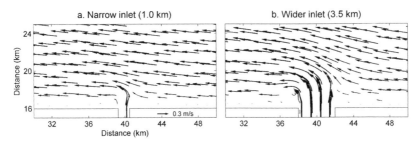

Figure 3-28 Depth averaged velocity vectors at mid-ebb water level (period 4 in Figure 3-11)

Initial inlet width significantly influences the evolution of inlet cross-section (Figure 3-29). Widening of the inlet cross-section occurs due to the erodible banks in the inlet gorge. The narrow inlet continuously widens because of the relatively strong velocities. However, the wider inlet results in lower velocities leading to slower evolution. It is evident in both models that the rate of widening decreases during the evolution as cross-sectional area increases. Further, the wider inlet becomes more or less stable at the end of the 50 years morphological period. In contrast, the narrow inlet still shows a significant rate of widening at the end of the 50 years simulation period. It is likely that the narrow inlet reaches a similar cross-section as if the simulation were continued for longer.

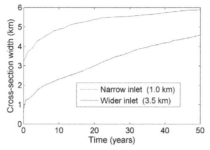

Figure 3-29 Temporal evolution of inlet cross-section width for narrow and wider inlets

The 50-year bed evolution shows significant differences in the predicted channel patterns (Figure 3-30). However, the predicted morphologies are consistent with measured Ameland inlet bathymetry. The narrow inlet results in higher velocities during flood- and ebb-phases. Therefore, the interaction between inlet and alongshore tidal currents results in higher velocities to the west of the inlet when compared to the wider inlet. This results in the evolution of the main inlet channel adjacent to the west barrier island at the seaward end and adjacent to the east barrier island at the landward end. The main channel gradually develops and shoal areas are formed to the east of the basin. This results in bifurcation of the main channel and the southward oriented branch becomes pronounced. However, the tidal currents still have the momentum to the east because of the propagation direction. Ultimately, an eastward oriented channel pattern is formed in the basin. In contrast, the wider inlet results in lower velocities at the west barrier island. Due to this, there is no strong orientation of the main channel. Instead, north-south oriented tidal currents are dominant in the inlet. This is evidence by the large seaward extension of the ebb-tidal delta and more sediment infilling in the basin. However, the northward extension of the channel is hampered due to the formation of ebb shoals. Thereafter, the main inlet channel is gradually oriented westward. This in turn results in an eastward oriented basin channel pattern. However, the channel pattern of the narrow inlet better resembles the typical morphological features found at the Ameland inlet.

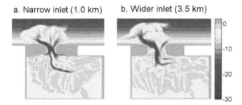

Figure 3-30 Effect of inlet width on the 50-year bed evolution; Narrow inlet (a), Wider inlet (b)

The hydrodynamic patterns are governed by the bed configurations (Figure 3-31). During the mid-flood, the narrow inlet results in higher flood velocities compared to the wider inlet case. This is probably due to the strongly developed westward main channel in the narrow inlet. Further, the results show that the presence of ebb-tidal delta enhances weaker currents east of the inlet. The alongshore currents are deflected resulting in strong bypassing currents in front of the ebb-tidal delta which acts as a barrier for the alongshore currents. These phenomena are accelerated in the case of wider inlet due to the larger extension of the ebb-tidal delta. These observations are in agreement with the hypothesis of Sha (1989).

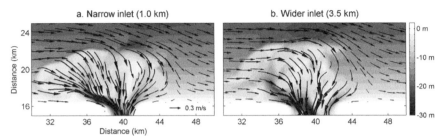

Figure 3-31 Interaction of alongshore tidal currents and ebb-tidal delta; Narrow inlet (a), Wider inlet (b)

3.6.2.3   Basin location

Three model domains were constructed with slightly different basin locations (Series 3 in Table 3-7). In the first, the back barrier basin is asymmetric to the east similar to the present orientation of the Ameland inlet. The second considers a symmetrical position of the basin relative to the inlet. The third consists of a westward asymmetric basin.

The predicted bed evolutions after 50 years for the 3 cases are shown in Figure 3-32. All model domains result in a westward oriented main inlet channel and ebb-tidal delta. However, the basin channel pattern shows different configurations in each case. When the basin is located to the west (Figure 3-32*a*), a two-channel system is found in the inlet. Both channels begin west of the seaward inlet end and develop along the west and east barrier islands. The channel along the east barrier island occurs due to the propagation direction of tide. The other channel (along the west island) is formed due to westward asymmetry of the basin. The symmetrical basin location results in a more symmetrical basin channel pattern (Figure 3-32*b*). The main channel in the basin is not pronounced in the east because of the symmetrical basin location. However, the eastward channels appear to be marginally pronounced than the westward channels due to the propagation direction of tide. As discussed earlier (see *section 3.6.2.2*), the eastern basin channel pattern is more pronounced when the basin is asymmetric to the east (Figure 3-32*c*). These morphological patterns agree with the sketches of Van Veen (1936). The channel pattern of a basin is eastward oriented, when the basin coast is straight and tide propagates from west to east (e.g. Figure 3-32*c*). This is due to the fact that the basin location is in favour to the tidal propagation direction and the eastern tidal prism is larger than the western.

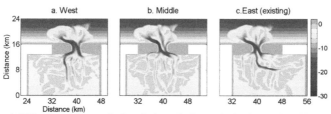

Figure 3-32 Predicted 50-year bed evolution relative to the basin location; West (a), Middle (b) and East-existing (c)

Figure 3-33 shows the temporal evolution of the ebb-tidal delta area which encloses terminal lobe and shoreline (Gibeaut and Davis, 1993). The initial rapid rate of growth decreases when the channel patterns are developed. The ebb-tidal delta area appears to be similar in all model domains (i.e. 52 (west), 48 (middle) and 50 km^2 (east)). The marginal difference can be described in terms of the seaward sediment transport and the characteristics of the ebb-tidal currents in the inlet which differ with the basin location. The ebb currents are not so strong when the basin is symmetrically located (Figure 3-32b). This results in the lowest seaward transport leading to the lowest ebb-tidal delta area. In the other two cases, strong ebb currents occur in the inlet due to the asymmetrical basin location. The ebb currents of the eastward basin are aligned with the tidal propagation (Figure 3-32c). In contrast, the westward basin results in a relatively disturbed current pattern in the inlet gorge (i.e. changing the course) which results in a different transport pattern (Figure 3-32a). This probably causes increased seaward transport leading to a marginal increase of the ebb-tidal delta area in the case of the western basin.

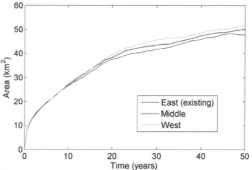

Figure 3-33 Temporal evolution of the ebb-tidal delta area relative to the basin location

3.7   Conclusions

Numerical experiments were undertaken with the process-based model (Delft3D) at decadal time scales to investigate the morphological evolution of a schematised tidal inlet. The schematised inlet represents the physical/hydrodynamic characteristics of the Ameland inlet which consists of typical large inlet/basin morphological patterns (i.e. eastward oriented basin channel system, westward oriented ebb-tidal delta and main inlet channel). Short-term simulations were undertaken to investigate hydrodynamic patterns of the inlet while long-term morphodynamic simulations qualitatively investigated the typical morphological patterns.

Short-term simulations re-produced the hydrodynamic behaviour of the study area reasonably well. This serves the purpose of this analysis. The results agreed with previous conceptual hypotheses that describe the water motion around the inlet. Alongshore and inlet tidal currents reached maximum value at mid-tide resulting in two dominant periods which affect the inlet morphology. M_2 velocity ellipses and residual flow patterns indicated the weaker and rotational character of tidal currents to the east of the inlet. The residual transport pattern indicated the potential to develop two inlet channels.

The analysis resulted in the determination of optimal model parameters (i.e. *MORFAC*, *dry cell factors* and *bed slope factors*) for the present model configuration. The VR transport formula predicted more prominent and diffused morphological evolution relative to the EH formula. The direction and asymmetry of tidal forcing strongly defined the inlet morphology implying that it is of utmost importance to take the dominant tidal characteristics of the study area into account. Small perturbations on the initial bathymetry resulted in different channel/shoal patterns. However, the typical morphological features were not significantly affected by these perturbations. Initial inlet width affected the bed evolution significantly. An initially narrow inlet resulted in channel/shoal patterns that better resembled present day conditions at the Ameland inlet. The effect of the basin location was marginal on the ebb-tidal delta evolution while that had a significant influence on the basin channel pattern.

This analysis suggests that the schematised process based modelling approach is capable of developing morphology that is representative of the present day morphology at the Ameland inlet (*BSS* > 0.3). This provides confidence in the ability of this approach to simulate large inlet/basin evolution at decadal time scales.

Chapter 4

Morphological response of tidal inlets to Relative Sea Level Rise

Much of the material of this chapter is based on,
Dissanayake, D.M.P.K., Roelvink, J.A. and Van der Wegen, M., 2008. Effect of sea level rise on inlet morphology, 7th International Conference on Coastal and Port Engineering in Developing Countries Dubai, United Arab Emirates, CD Rom.
Dissanayake, D.M.P.K., Ranasinghe. R., and Roelvink, J.A., 2009. Effect of sea level rise in inlet evolution: a numerical modelling approach, Journal of Coastal Research, SI 56, Proc. of the 10th International Coastal Symposium, Lisbon, Portugal, pp. 942- 946.
Dissanayake, D.M.P.K., Ranasinghe, R., Roelvink, J.A., (in press). The morphological response of large tidal inlet/basin systems to Relative Sea Level rise, Climatic Change.

4.1   Introduction

Large tidal inlet/basin systems, which usually contain extensive tidal flats that are rich in bio-diversity, are commonly found along the coasts of the Netherlands (Sha, 1989), USA (Fenster and Dolan, 1996), China (Gao and Jia, 2003), Bangladesh (Ortiz, 1994), and Vietnam (Duc, 2008). The rich bio-diversity in these systems results in their localities becoming major tourist attractions, thus generating billions of tourism dollars for the local economy. Due to the associated increase in economic activities, local communities in these areas have grown rapidly in recent decades. The continued existence and/or growth of these environmental systems and communities are directly linked to the extensive tidal flats in the basins that are host to a plethora of diverse flora and fauna. However, these tidal flats are particularly vulnerable to any rise in the mean sea level and may be very sensitive to potential sea level rise (SLR) impacts such as coastline transgression (landward retreat), regression (seaward advance), erosion of ebb-tidal deltas and sedimentation of tidal basins (Dissanayake et al., 2009b). In view of projected climate change impacts, a clear understanding of the potential impacts of relative sea level rise (i.e. Eustatic SLR and local effects such as subsidence/rebound etc.) on these inlet/basin systems is therefore a pre-requisite for the sustainable management of both the inlet/basin system and the communities that depend on them. To date however, little is known about the potential impact of relative sea level rise (RSLR) on this type of systems.

Only a few studies have investigated the potential impacts of RSLR on large inlet/basin systems. Friedrichs et al (1990) investigated the impact of RSLR on tidal basins on the US East coast using the simple one-dimensional numerical model of Speer and Aubrey (1985) and found that RSLR results in sediment import or export depending on local basin geometry. Van Dongeren and De Vriend (1994) investigated the RSLR induced morphological behaviour of the Frisian inlet in the Dutch Wadden Sea using a one-dimensional semi-empirical model with a time-invariant RSLR. Using historical bathymetric surveys, Louters and Gerritsen (1995) found that the Wadden Sea accumulates sediment and bed levels keep up with RSLR. The sediment balance in the wider North Sea basin for the Holocene sea level rise was modelled by Gerritsen and Berentsen (1998). However, their large-scale results are not directly applicable at single tidal basin scale. Dronkers (1998) analysed the net sediment transport behaviour of the Wadden Sea tidal basins using an analytical approach and suggested that the basins are generally flood dominant and that RSLR tends to result in sediment accumulation in order to restore the dynamic equilibrium of the basin. Van Goor et al (2003) adopted a semi-empirical modelling approach (ASMITA) to investigate the impact of RSLR on two of the Wadden Sea inlets, Ameland and Eierland, and showed the inlet/basin morphology of the two systems could keep up with a RSLR of more than 10 mm/year and 15 mm/year respectively.

The major shortcoming of the above studies is that none of them provided useful insight into the RSLR induced morphological changes in the inlet/basin morphology at spatial resolutions that are useful for efficient environmental and coastal management/planning in these socio-economically and environmentally sensitive areas. The study presented herein, which focuses on the Ameland inlet in the Dutch Wadden Sea area, attempts to address this knowledge gap by investigating the morphological evolution of a typical large inlet/basin system in response to RSLR at relatively high temporal and spatial resolutions.

4.2   Modelling philosophy

The traditional morphodynamic modelling philosophy is to calibrate and validate a model using laboratory and/or field measurements and then use the calibrated/validated model in hindcast or forecast mode to obtain quantitative estimates of system response to forcing. Depending on model accuracy, quality of data, and the accuracy of forcing conditions used for the hindcast/forecast, this traditional approach (i.e. virtual reality) may provide quantitatively accurate morphological predictions over relatively short time scales (~days-months). The present application is, however, a qualitative long term morphodynamic modelling exercise (~ 100 years) for which this traditional 'virtual reality' approach is not ideally suited due to several reasons. First, it is well known that a calibrated/validated model is quite likely to depart from the 'truth' the farther the simulation progresses past the calibration/validation period (Figure 4-1). This is largely due to the almost unavoidable initial imbalance between initial model bathymetry and hydrodynamic forcing resulting in large and unrealistic morphological changes at the beginning of the simulation while the model attempts to create a morphology that is more or less in equilibrium with the forcing. Depending on model settings, these initially large changes may result in a positive feedback loop which may eventually produce nonsensical predictions. The deterministically chaotic nature of numerical models (Lorenz, 1972) is another phenomenon that places significant uncertainties on the quantitative accuracy of long term morphodynamic model predictions. Secondly, significant input reduction is required for long term simulations (i.e. present computational costs do not allow multiple long term brute force simulations). Therefore, even a model that is perfectly calibrated/validated (i.e. representing real forcing/response relationships), cannot be expected to predict future system response accurately due to the highly schematised nature of future forcing that is unavoidable in long term morphodynamic simulations. Finally, to have confidence in the 'virtual reality' approach, the calibration/validation periods should be of a similar duration to the hindcast/forecast period. As such, to have a satisfactory level of confidence in a 100 year simulation, ample high quality calibration/validation data should be available over the last 100 years. This is clearly not the case as older (pre 1950) morphology/hydrodynamic data is of questionable quality, and also very sparse.

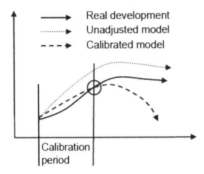

Figure 4-1 Pessimistic scenario for effect of calibration: real development, unadjusted model and calibrated model (Virtual reality) (after Roelvink and Reniers, 2011)

For the long term forecasts necessary in this study, therefore, a different modelling philosophy ('Realistic analogue' (Roelvink and Reniers, 2011)) is adopted. The 'Realistic analogue' philosophy essentially commences the simulation with a highly schematised initial bathymetry and allows the model to gradually produce the morphology that is in equilibrium with the main forcing (e.g. tidal forcing) that is to be used in forecast mode. The level of schematisation in the initial bathymetry is extreme in that while the geometry of the coastal system under investigation is very broadly represented, the initial bed is assumed to be more or less flat (i.e. flat bed morphology). Once this initial 'establishment simulation' produces a near-equilibrium morphology that is sufficiently similar to the observed system in terms of channel/shoal patterns and typical morphometric properties, the simulation can be extended into the future with slightly varied forcing (e.g. tidal forcing and slow sea level rise) to qualitatively investigate future system behaviour. It is, however, crucial, that the model generated 'equilibrium morphology' be compared with data to ensure that it is indeed an equilibrium morphology. This requirement, does limit the application of the 'Realistic analogue' philosophy to mature systems that are currently in equilibrium or have previously been in equilibrium prior to human interventions. It is emphasised that the 'Realistic analogue' approach is only suitable for qualitative assessments of long term system response to forcing.

4.3   Establishment simulation

In this study the "Realistic analogue" philosophy described in the previous section is adopted. The first step then is to analyse existing bathymetric data to determine whether the study area is in (or close to) morphological equilibrium. Two equilibrium relations (tidal prism vs inlet cross-sectional area (Jarret, 1976; Eysink, 1990); and tidal amplitude to mean channel depth ratio vs shoal volume to channel volume ratio (Wang et al., 1999), see *section* 4.6.3) were used for this purpose. Bathymetric data from 1930 to 2004 were available for the analysis. Figure 4-2 shows the bathymetry data in comparison to the equilibrium relations. Arrows indicate the general trend of the data points. If the points are closer to the lines, they tend to be equilibrium. The points on the lines imply equilibrium systems. According to both empirical equilibrium relations, the 2004 bathymetry appears to be very close to equilibrium conditions (Figure 4-2). Therefore, the "Realistic analogue" philosophy then dictates that the initial 'establishment simulation' starting from a schematised flat bed bathymetry should reproduce a bathymetry that is qualitatively similar to the 2004 measured bathymetry. An establishment simulation was thus undertaken, starting with the schematised bathymetry (see Figure 3-7), forced with M_2, M_4 and M_6 tidal constituents.

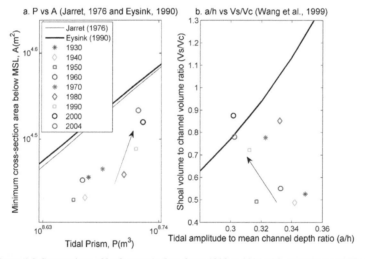

Figure 4-2 Comparison of bathymetric data from 1930 to 2004 with empirical equilibrium relations, (a) Tidal prism vs inlet cross-sectional area (Jarret, 1976 and Eysink, 1990), (b) Tidal amplitude to mean channel depth ratio vs shoal volume to channel volume ratio (Wang et al., 1999)

At the start of the 'Establishment simulation' an inlet width of 1 km was specified in line with historical observations (Rijzewijk, 1981). Two separate simulations were undertaken: one with a single *MORFAC* of 200 for the entire simulation and the other with time varying *MORFAC*s where smaller *MORFAC*s gradually changed to larger *MORFAC*s as the simulation progressed in time (i.e. *MORFAC* = 10 for 0 – 5 years, *MORFAC* = 20 for 5 – 15 years, *MORFAC* = 50 for 15 – 30 years and *MORFAC* = 200 for 30 – 50 years). A *MORFAC* value of 200 was selected for the single *MORFAC* simulation following the results of rigorous performance comparisons of simulations with *MORFAC*s of 50, 100, 200 and 400 (see *section 3.6.1*).

A very rapid morphological evolution is shown at the beginning of both simulations while the initial flat bed bathymetry attempts to reach a state that is in equilibrium with the forcing (see, for example, the growth rate of the ebb-tidal delta during the first 20 years of the single *MORFAC* simulation shown in Figure 4-4*a*). However, after this initial rapid adjustment, the evolution of both the ebb-tidal delta and the basin appears to reach quasi-steady conditions after 50 years of simulation (see Figure 4-4) implying near-equilibrium conditions. Therefore, the length of the establishment simulation, the purpose of which is to establish a morphology that is in (or close to) equilibrium, was limited to 50 years.

4.3.1   Visual comparison

Figure 4-3 shows the model predicted morphologies at the end of the 50 years simulation period for both cases, is shown in comparison with the 2004 measured bathymetry. A visual comparison indicates that both model predictions are fairly consistent with the measurements as far as the main morphological features are concerned. Both the measured and modelled morphologies consist of an inlet with a two-channel system. Inside the basin, the main inlet channel extends towards the east, while a secondary channel extends towards the west. One major difference between the modelled and measured morphologies is that, in the measurements, the main inlet channel is located adjacent to the eastern barrier island while in the modelled morphologies it is located close to the western barrier island. The ebb-tidal delta and the main inlet channel are skewed to the west in both the measurements and model predictions. However, the western part of the model predicted ebb-tidal delta consists of more shoal areas in comparison to the measurements. The observed differences between the model predicted and measured morphologies are likely to be due to a) the highly schematised initial bathymetry, and b) the non inclusion of wave effects (see *Chapter 5*).

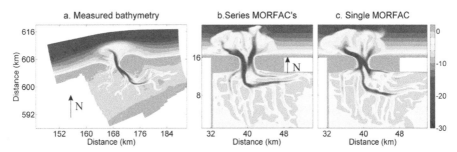

Figure 4-3 Comparison of modelled and measured morphologies; 2004 Ameland measured morphology (a), predicted morphology with time varying *MORFAC*s (b) and single *MORFAC* (c)

4.3.2   Statistical comparison

To quantitatively determine whether the single or time varying *MORFAC* approach gives a better representation of the present bathymetry, a statistical analysis was undertaken using the Brier Skill Score (*BSS*) (Sutherland et al., 2004; Van Rijn et al, 2003). It is noted that the *BSS* is not a perfect method to evaluate model performance especially where individual characteristics such as basin infilling, lateral displacement and rate of bifurcation of channels etc. are concerned. However, at this point in time this is the most widely adopted method to objectively assess model skill (Ruessink et al., 2003; Pedrozo-Acuna at al., 2006; Ruggiero et al., 2009; Roelvink et al., 2009), and due to the lack of a better alternative, the *BSS* approach is adopted to evaluate model skill. The *BSS* is defined by Van Rijn et al (2003) as discussed in *section 3.6.1* (*Eq.* 3-19).

A more comprehensive definition of the *BSS* is given by Sutherland et al (2004):

$$BSS = \frac{\alpha - \beta - \gamma + \varepsilon}{1 + \varepsilon} \qquad\qquad \text{4-1}$$

where, $\alpha = r^2_{YX'}$ $\beta = \left(r_{YX'} - \frac{\sigma_{Y'}}{\sigma_{X'}} \right)^2$ $\gamma = \left(\frac{\langle Y' \rangle - \langle X' \rangle}{\sigma_{X'}} \right)^2$ $\varepsilon = \left(\frac{\langle X' \rangle}{\sigma_{X'}} \right)^2$

and $Y' = z_{measured} - z_{Flat}$ $X' = z_{MORFAC} - z_{Flat}$

α, a measure of bed form phase error and perfect model gives $\alpha = 1$
β, a measure of bed form amplitude error and perfect model gives $\beta = 0$
γ, an average bed level error and perfect model gives $\gamma = 0$
ε, a normalization term which indicates the measurement error

Table 4-1 indicates the classification of Sutherland et al (2004) for the assessment of the model performance against the *BSS*.

Classification	BSS$_{Sutherland}$
Excellent	1.0 – 0.5
Good	0.5 – 0.2
Reasonable/Fair	0.2 – 0.1
Poor	0.1 – 0.0
Bad	<0.0

Table 4-1 Classification for Brier Skill Score by Sutherland et al (2004)

In the present application, *BSS* values for the two simulations were calculated using both methods described above for a control area which contains the area of interest consisting of the ebb-tidal delta, inlet and basin areas (Figure 3-16c). Table 4-2 shows decomposition terms (*α, β, γ* and *ε*) of Sutherland et al (2004) and the estimated *BSS* values of both methods. Difference in phase (*α*) and amplitude (*β*) of the two simulations is significant compared to that of mean (γ) and normalisation term (ε) indicating that phase and amplitude dominate on the *BSS* value. The *BSS* values thus calculated (Table 4-2) indicate that, according to Van Rijn et al (2003), the comparison between measurements and morphologies predicted with single and time varying *MORFAC* approaches are reasonably good and poor respectively while both predicted morphologies are good according to Sutherland et al (2004) while the single *MORFAC* simulation results in a higher *BSS* value compared to the time varying *MORFAC* simulation. Thus, the morphology predicted by the single *MORFAC* approach after a 50-year simulation period appears to be consistently and sufficiently close to the equilibrium morphology at the study area. Following the "Realistic analogue" philosophy, this equilibrium bathymetry is therefore adopted as the initial bathymetry to qualitatively investigate system response to various RSLR scenarios in subsequent model simulations. For convenience, this bathymetry (shown in Figure 4-3*c*) will be referred to as 'initial bathymetry' from hereon.

Bathymetry	BSS$_{Van\ Rijn}$	BSS$_{Sutherland}$				
		α	β	γ	ε	BSS
Single *MORFAC*	0.35	0.45	0.02	0.03	0.12	0.47
Time varying *MORFAC*	0.20	0.32	0.06	0.02	0.10	0.31

Table 4-2 BSS values for the two different *MORFAC* approaches according to Van Rijn et al (2003) and Sutherland et al (2004)

4.3.3   Ebb-tidal delta and tidal basin evolution

The evolution of the ebb-tidal delta and tidal basin volumes during the Establishment simulation is shown in Figure 4-4.

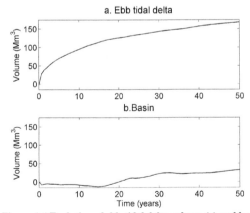

Figure 4-4 Evolution of ebb-tidal delta volume (a) and basin volume (b) in single *MORFAC* approach

The ebb-tidal delta volume is estimated relative to the undisturbed bathymetry which consists of shore parallel contours (Walton and Adams, 1976). The basin volume is defined as the sediment volume with reference to the initial flat bed. For the first 15 years of the Establishment simulation, the basin volume is slightly negative and then turns positive as a result of pattern formation on the flat bed. The ebb-tidal delta volume changes from zero to about 180 Mm^3 by the end of the Establishment simulation, which is, to a first order, comparable, to the 2004 measured ebb-tidal delta volume of about 130 Mm^3 (see *section 3.6.1*). Figure 4-4 indicates an initially strongly ebb dominant system which becomes weakly flood dominant by the end of the 50 years simulation. This is consistent with contemporary observations of flood dominance at the study site (Dronkers, 1986; Ridderinkhof, 1988; Dronkers, 1998). These observations further justify the adoption of the morphology predicted at the end of the establishment simulation as a proxy for the present day morphology of the study site.

4.4   RSLR scenarios

Present analysis used two RSLR categories viz. time-invariant (i.e. constant rate of RSLR) and time-variant (i.e. variable rate of RSLR) consisting three scenarios in each category (Table 4-3 and Table 4-4).

4.4.1   Time-invariant RSLR scenarios

Table 4-3 shows applied three time-invariant RSLR scenarios. The second scenario (i.e. 5 mm/year) is based on Van Dongeren and De Vriend (1994). Accordingly, the expected sea level increase is about 0.4 to 0.6 m in the next 100 years in the study area. The third scenario (i.e. 10 mm/year) is adopted following the semi-empirical modelling approach of Van Goor et al (2003). Three simulations were undertaken for 50 years starting from the initial bathymetry.

No	RSLR Scenario	Description
1	No RSLR	MSL is constant throughout the model duration
2	5 mm/year	Averaged RSLR in the study area for the next 100 years (Van Dongeren and De Vriend, 1994)
3	10 mm/year	Predicted critical RSLR in the study area based on the semi-empirical approach of Van Goor et al (2003).

Table 4-3 Time-invariant RSLR scenarios to investigate the initial bathymetry evolution for 50 years

4.4.2   Time-variant RSLR scenarios

Time-variant RSLR consists of two main components: Eustatic SLR and local vertical land movement (see *section 2.2*). Projections for global average Eustatic SLR are given in IPCC (2001, 2007). In this study, only the lower and higher ensemble average SLR projections (i.e. 0.2 m and 0.7 m by 2100 compared to 1990) are taken into account. Vertical land movement is a result of subsidence and glacial rebounds. A local subsidence due to gas extraction has been observed in the study area (Marquenie and Vlas, 2005), and it is expected to be between 0 – 0.1 m over the next 50 years (Van der Meij and Minemma, 1999). In the simulations undertaken herein, a land subsidence of 0.1 m in the next 50 years was specified, representing the worst case scenario. The effect of land subsidence is introduced by means of raising the mean water level in the model domain. Table 4-4 shows three scenarios investigated for 110 years evolution starting from the initial bathymetry.

No	RSLR scenario	Description
1	No RSLR	MSL is constant throughout the model duration
2	IPCC L + LS	The combined effect of IPCC lower ensemble average projection (0.2 m rise in MSL by 2100 compared to 1990) and land subsidence (0.15 m*)
3	IPCC H + LS	The combined effect of IPCC higher ensemble average projection (0.7 m rise in MSL by 2100 compared to 1990) and land subsidence (0.15 m*)

Table 4-4 Time-variant RSLR scenarios to investigate the initial bathymetry evolution for 110 years
* The rate of local land subsidence is expected to be 0-0.1 m over the next 50 years and decrease thereafter (Van der Meij and Minemma, 1999). Thus, a maximum land subsidence value of 0.15 m over the 110 years (1990-2100) is specified in the scenario 2 and 3, which is expected to represent the worst case scenario.

The time-variant RSLR scenarios were implemented in the simulations as a part of long period harmonic waves (i.e. wave length/4). The Eustatic SLR appears to be accelerating while the rate of land subsidence decreases as the consolidation occurs. Therefore, two harmonic signals were formulated for these phenomena.

From *Eq.*   3-17, time-varying and spatially uniform harmonic signal reads as,

$$\eta(t) = \hat{\eta}\cos(\omega t - \varphi)$$

4-2

The increase of MSL due to RSLR was considered as the amplitude ($\hat{\eta}$) and the period of harmonic wave was four times of the hydrodynamic period resulting to an angular frequency of $2\pi/(4 \times Hydrodynamic\ period)$. Phase lag ($\varphi$) corresponds to the RSLR phenomenon (i.e. 180^0 and 90^0 represents the Eustatic SLR and land subsidence respectively).

Figure 4-5a shows the application of the scenario 2 and 3 via the harmonic schematisation. As per projections, the rate of Eustatic SLR increases during most of the 110 years period while the rate of land subsidence increases slightly for the first 50 years and then decreases (Figure 4-5b).

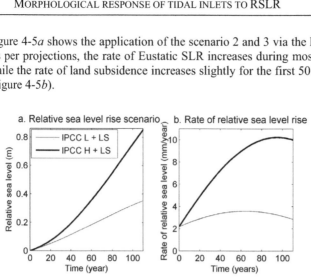

Figure 4-5 RSLR scenarios for 110-year period; magnitudes of RSLR (a) and rates of RSLR (b), for IPCC low projection and Land subsidence (IPCC L+LS), and IPCC high projection and Land subsidence (IPCC H+LS) scenarios

Applying these RSLR scenarios results in increasing the MSL of the model area. However, it is emphasised that the bed evolutions are hereon shown with respect to the initial MSL.

4.5   Morphological response to time-invariant RSLR

4.5.1   Bed evolutions

Figure 4-6 shows the 50-year bed evolutions of the time-invariant RSLR scenarios starting from the initial bathymetry. All predicted evolutions consist of the typical morphological features of the Ameland inlet. Results indicate erosion on the ebb-tidal delta and infilling in the basin. This is characterised by a flood dominant transport system in the inlet which strongly corresponds to the measured data of the Ameland inlet. Resulting evolution of the main inlet channel is adjacent to the west barrier island at the seaward inlet end and to the east barrier island at the landward inlet end due to the direction of tidal wave propagation. The bed evolution tends to have pronounced as the rate of RSLR increases whereas the No RSLR scenario indicates the southward orientation of the main inlet channel in the basin. The channel pattern suggests that strong eastward oriented tidal currents are resulted in the inlet gorge as the MSL increases.

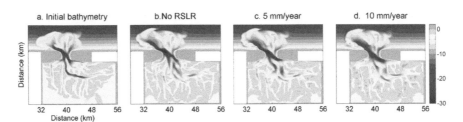

Figure 4-6 50-year evolutions of initial bathymetry with time-invariant RSLR scenarios; Initial bathymetry (a), No RSLR (b), 5 mm/year (c) and 10 mm/year (d)

Visual discern recommends no significant differences among the final predicted morphologies. Therefore, quantitative comparisons amongst the 3 RSLR scenarios were undertaken in terms of the ebb-tidal delta and basin evolution.

4.5.2   Ebb-tidal delta evolution

Figure 4-7 shows the evolution of the ebb-tidal delta area (see *section 3.6.2.3*) and volume (see *section 3.6.1*) of the time-invariant RSLR scenarios. In the first decade, applications of all scenarios indicate similar evolution. This is due to the marginal effect of the RSLR at the beginning. Thereafter, the No RSLR scenario shows gradual increasing and reaches more or less stable ebb-tidal delta area. This is evidence of comparatively low flood dominant condition. In contrast, the other two RSLR scenarios tend to develop lower ebb-tidal delta areas of which the lowest area is resulted in the highest RSLR scenario. This implies strong erosion on the ebb-tidal delta as the RSLR increases. In the last decades, the northwest-southeast oriented main inlet channel is pronounced in the non-zero RSLR scenarios. Thus, the strong ebb-tidal delta currents in the channel probably diffuse sediment on the ebb-tidal delta leading to increase the area (i.e. increase of No RSLR ~ 1 Km^2 and the highest RSLR ~ 3 km^2 in the last decade). The final predicted morphologies show the highest area in the No RSLR (~ 56 km^2) and the lowest area in the highest RSLR (~ 54 km^2). However, the measured area is about 70 km^2 of the Ameland inlet. This discrepancy is probably due to the schematised approach and non-inclusion of the wave effect (see *Chapter 5*).

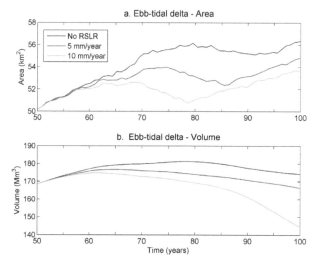

Figure 4-7 Ebb-tidal delta evolution with time-invariant RSLR scenarios starting from the initial bathymetry; area (a) and volume (b)

Initially, the evolution trend of the ebb-tidal delta volumes quite agrees with that of the areas (Figure 4-7b). Later, the RSLR effect gradually dominates leading to strong decrease in the ebb-tidal delta volume as the RSLR rate increases. In the last decades, all scenarios show decreasing the volume while increasing the area. This is due to the definitions of these parameters (i.e. area considers spatial extension whereas volume considers sediment amount). The No RSLR scenario resulted in the highest volume (~175 Mm^3) while the highest RSLR scenario indicates the lowest volume (~145 Mm^3). However, the measured ebb-delta volume is about 130 Mm^3 of the Ameland inlet.

4.5.3   Tidal basin evolution

Basin area-depth hypsometry

Figure 4-8 shows basin area-depth hypsometry based on the final predicted morphologies of the RSLR scenarios and the 2004 measured data of the Ameland inlet. The predicted bathymetries are deeper in the shallow areas (i.e. depth < 5 m) whereas the deep areas (i.e. depth > 10 m) quite agree with the data. This is expected due to deep channels on the predicted beds and undeveloped shallow areas in the basin (Figure 4-6). The characteristics of the tidal flat evolution are shown in the zoom out view. The predicted area evolution corresponds to the rate of RSLR such that the higher the rate of RSLR the larger the area. This further implies decreasing tidal flat area while increasing the rate of RSLR.

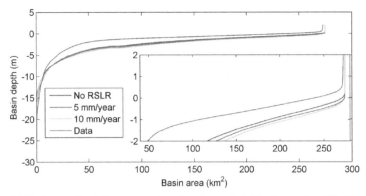

Figure 4-8 Basin area-depth hypsometry based on the predicted bathymetries of the RSLR scenarios and the 2004 measured data of the Ameland inlet

The hypsometry indicates the overall geometry of the basin and does not provide spatial extent of tidal flats/channels. Following analysis shows the spatial distribution of tidal flats/channels in the basin.

Spatial distribution of tidal flats and channels

The bed evolutions showed pronounced shoal areas in the basin as the rate of RSLR increases (Figure 4-6). However, these are based on a constant MSL whereas it increases with the rate of RSLR. Thus, the present analysis estimates the spatial distribution of tidal flats/channels on the predicted morphologies with increased MSL. In order to get a relative comparison between RSLR scenarios, the main channel areas are traced by 5 m depth contour while the flat areas are defined as the dry surface area below 1 m from MSL (i.e. no shoal areas are found beyond mean high water (MHW) and the tidal range is about 2 m). Figure 4-9 shows the spatial pattern of tidal flats (brown patches) and channels (blue lines) in the basin. The channel pattern agrees to the present-day bathymetry and that tends to expand as the rate of RSLR increases. In contrast, the flat areas gradually diminish and these are relatively concentrated to the east of the basin because of the eastward oriented basin channel pattern.

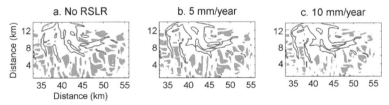

Figure 4-9 Spatial distribution of tidal flat (dry area below 1 m) and channel areas (5 m depth contour) on final predicted beds; No RSLR (a), 5 mm/year (b) and 10 mm/year (c)

Temporal evolution of tidal flat and channel volumes

Channel volume is the volume of water below mean low water (MLW) while tidal flat volume is the sand volume between MLW and MHW (Figure 4-17c). The temporal evolution of these two elements is shown in Figure 4-10. The evolutions of channel/tidal flat volumes have been significantly modulated by the RSLR. The initial volume corresponds to the initial bathymetry. In contrast to the ebb-tidal delta (Figure 4-7), the channels/tidal flats volumes have strong evolution from the beginning of the morphological period implying that the basin evolution is highly sensitive to the RSLR. The schematised approach consists of a confined basin area which results in decreasing the tidal flat volume while increasing the channel volume. The rate of volume change corresponds to the rate of RSLR such that the higher the RSLR the lower/higher tidal flats/channels volume. RSLR accelerates sediment import and creates positive accommodation space in the basin (Beets and Van der Speck, 2000; Coe and Church, 2003; Cowell et al., 2003; Zhang et al., 2004). Therefore, a stable tidal flat evolution is determined by counterbalancing of these two phenomena. The No RSLR results in increasing tidal flats and decreasing channels volume due to the sediment import into the basin with constant MSL. These are evidence of a flood dominant system (Friedrichs and Aubrey, 1988; Dronkers, 1998). In case of the highest RSLR, the tidal flat volume decreases implying lower sediment import into the basin compared to the accommodation space created by increased MSL. However, the low RSLR scenario (i.e. 5 mm/year) tends to indicate more or less stable tidal flat evolution. Thus, it appears to be the critical rate of RSLR of the system.

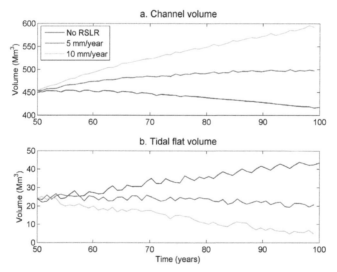

Figure 4-10 Temporal evolution of tidal basin with time-invariant RSLR scenarios; channel (a) and tidal flats (b)

4.6   Morphological response to time-variant RSLR

4.6.1   Bed evolution

Visual Comparison

Figure 4-11 shows the predicted morphology at the end of the three 110 year simulations (Table 4-4). In general, two major differences can be identified in these patterns with respect to the initial bathymetry; eroding of the ebb-tidal delta, and significant infilling of the basin. These characteristics become more pronounced with increasing RSLR.

The shoal areas of the model predicted final ebb-tidal deltas are generally smaller than that of the initial bathymetry. This is largely due to the formation of pronounced channels within the ebb-tidal delta. More ebb-tidal delta channels are formed as the RSLR is increased leading to the smallest shoal area in the Scenario 3 (i.e. IPCC H+LS) (Figure 4-11*d*). This implies accelerated erosion of the ebb-tidal delta due to the RSLR.

In the two non-zero RSLR scenarios (Figure 4-11*c* and *d*), significant infilling of the basin leads to the disappearance of the pronounced channel configuration that is initially present in the basin. The shoal area (relative to initial MSL) increases as the RSLR increases. However, the dominant northwest-southeast orientation, following the propagation direction of tidal wave (see *section 3.6.2.3*), is still apparent at the end of all three RSLR scenario simulations.

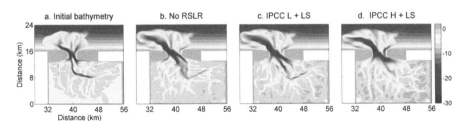

Figure 4-11 Initial bathymetry (at the end of the 50 year Establishment simulation) (a), and final model predicted bathymetry after a 110-year period for No RSLR (b), IPCC Lower prediction and land subsidence (c) and IPCC Higher prediction and land subsidence (d)

Statistical comparison

Here, the model predicted bed level changes are quantified using two statistical methods. This analysis was also undertaken using the control area shown in Figure 3-16*c*. The two statistical parameters employed are (a) the correlation coefficient (R^2), and (b) the *BSS* value of Van Rijn et al (2003). Both parameters are used to quantify the final predicted bed level changes (with respect to the initial bathymetry) between the RSLR and 'No RSLR' Scenarios.

Bed level changes (and not bed levels) are used in the calculation of the R^2 to ensure that areas of no change (e.g. barrier islands) are not included in the calculation. This ensures that there is no bias towards high R^2 values due to the presence of such 'no change' areas inherent in the bathymetry.

R^2 is defined as:

$$R^2 = 1 - \frac{\sum(\Delta z_{NoRSLR} - \Delta z_{Scenario})^2}{\sum(\Delta z_{NoRSLR} - \langle \Delta z_{NoRSLR}\rangle)^2}$$

4-3

where, Δz_{NoRSLR} is the bed level change of the No RSLR scenario with respect to the initial bathymetry. Similarly, $\Delta z_{Scenario}$ is the relative bed level change for the low and high RSLR scenarios.

High R^2 values indicate a high degree similarity between the bed level changes in the RSLR and No RSLR scenarios, and thus no significant morphological influence of the RSLR. Figure 4-12 shows the results of this comparison. The solid line indicates the linear fit and the cloud shows the density of the bed level changes between two data sets. The darker the cloud area the higher the density. Generally, most bed level changes are found in the range of 0 – 10 m in depth. Scenario 3 (IPCC H + LS) is associated with the lowest correlation coefficient (Figure 4-12*b*) implying the highest bed level changes compared to the No RSLR scenario.

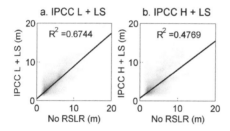

Figure 4-12 Correlation between bed level changes (with respect to the initial bathymetry) between RSLR and No RSLR scenarios. The solid line indicates the correlation and the cloud shows the density of bed level changes; Low RSLR (a) and High RSLR (b) scenarios

The second statistical parameter used here is the *BSS* value, given by:

$$BSS = 1 - \frac{\langle(z_{Scenario} - z_{NoRSLR})^2\rangle}{\langle(z_{Initial} - z_{NoRSLR})^2\rangle}$$

4-4

where, $z_{Scenario}$ and z_{NoRSLR} are the predicted bed levels of the RSLR scenarios (low and high) and the No RSLR scenario respectively. $z_{Initial}$ represents the initial bathymetry.
This definition of BSS provides a relative measure of the RSLR effect with respect to the No RSLR situation. As the denominator remains unchanged for both RSLR scenarios, the

BSS is determined by the numerator. As the numerator increases with $z_{Scenario}$, the BSS decreases with increasing $z_{Scenario}$.

Similar to the R^2, the *BSS* of 1 predicts excellent comparison between the bed level changes, implying in this case, an insignificant morphological influence of the RSLR. The calculated *BSS* values given in Table 4-5 and indicate that higher bed level changes are associated with Scenario 3 (IPCC H + LS), which is consistent with the above correlation analysis.

RSLR scenario	BSS
IPCC L + LS	0.69
IPCC H + LS	0.50

Table 4-5 BSS values for bed level changes (with respect to the initial bathymetry) between RSLR and No RSLR scenarios.

The above discussion qualitatively indicates that:
- *The bed level changes increase as the RSLR increases*
- *The ebb-tidal delta erodes as the RSLR increases*
- *The basin accretes as the RSLR increases*

All of the above phenomena can be, to first order, explained by the cumulative sediment transport through the inlet. Figure 4-13 shows the cumulative sediment transport through the middle cross-section of the inlet for the 3 Scenarios given in Table 4-4. Mildly flood dominant transport patterns are predicted for all 3 scenarios during the first 10 years implying quasi-stable conditions. This is probably because the RSLR is negligible during this period. After the first 10 years, the degree of flood dominance increases rapidly in time and increases with RSLR. This is probably because the tidal wave distortion due to inlet geometry and the reflected wave from the basin (Dronkers, 1986; Dronkers, 1998), both of which result in flood dominance, are enhanced due to RSLR. Higher RSLR will likely result in more enhancement of the tidal wave distortion, thus leading to the unsurprising result (shown in Figure 4-13) that flood dominance increases with increasing RSLR.

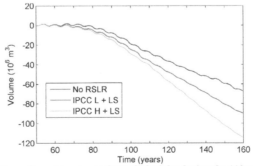

Figure 4-13 Cumulative sediment transport through the inlet during the 110-year simulation period for the three RSLR scenarios given in Table 4-4. Negative transport values imply landward (flood) transport

4.6.2   Evolution of morphological elements

Here the temporal evolution of the various morphological elements in the inlet/basin system under the three RSLR scenarios is described.

(a)   Ebb-tidal delta and basin volume

Figure 4-14 shows the variation of the ebb-tidal delta and basin volume during the 110 years simulations for the three different RSLR scenarios considered.

The ebb-tidal delta volume appears not to be correlated with the rate of RSLR during the first 20 years, probably due to the low rate of RSLR during this initial period (see Figure 4-5). After this period, the eroded ebb-tidal delta volume appears to be strongly positively correlated with the rate of RSLR with the No RSLR and IPCC H+LS resulting in the highest and lowest ebb-tidal delta volumes respectively. Analysis of tidal prisms, tidal asymmetry (M_2/M_4) and phase difference ($2M_2$-M_4) suggests that this correlation is not due to any RSLR induced increase in tidal prism but due to the RSLR driven enhancement of the tidal wave interaction with the inlet geometry and changes in the reflected tidal wave which results in increased flood dominance (Friedrichs and Aubrey, 1988). The degree of flood dominance is directly correlated with the rate of RSLR, thus causing the ebb-tidal delta erosion volume to increase with increasing rates of RSLR.

The basin volume change is inversely proportional to that of the ebb-tidal delta. The higher the erosion of the ebb-tidal delta the higher the sedimentation of the basin. The maximum volume decrease (Scenario 3) in the ebb-tidal delta is about 100 Mm^3. In contrast, the corresponding volume increase in the basin is about 200 Mm^3. This implies that there are additional sediment sources which are likely to include adjacent coastlines and erodible inlet banks. Model results imply that the volume of sediment supplied to the basin by these additional sources is of the same order of magnitude as that supplied by the ebb-tidal delta. The higher volume change in the basin (e.g. 127 Mm^3 higher than ebb-tidal delta volume change in Scenario 3) also implies that the basin is likely to be more vulnerable to rising MSL than the ebb-tidal delta.

It is emphasised however that the above observations do not take into account any wave effects on sediment transport. Simulations that incorporate wave forcing predict that the inclusion of wave effects generally results in less pronounced morphological elements (see *Chapter 5*). This is mainly due to wave-current interaction in the inlet gorge which augments flood-transport and reduces ebb-transport. Inclusion of wave effects also appears to result in relatively higher sediment distribution in the basin and the ebb-delta areas.

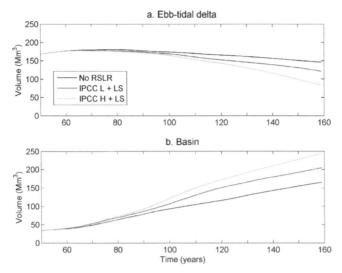

Figure 4-14 Evolution of the ebb-tidal delta volume (a), and the basin volume (b) during the 110 years simulation period for the three RSLR scenarios given in Table 4-4

(b)   Channel and tidal flat volume in the basin

The channel volume is almost constant during the first decade and slightly decreases thereafter under the No RSLR scenario (Figure 4-15a). The slight decrease in the channel volume after the first decade is probably due to the expansion of tidal flats resulting from substantial sediment infilling into the basin. In Scenario 2 (low RSLR), the channel volume increases only marginally (from 609 to 739 Mm^3) during the 110-year simulation. In contrast, a significant volume change (from 609 to 974 Mm^3) is predicted in Scenario 3 (high RSLR). This is because, under the latter scenario, the volume of sediment imported into the basin is not sufficient to fill the new accommodation space gained due to large rise in MSL.

The tidal flat volume depends on the channel slopes, the mean sea level and the sediment import into the basin (Friedrichs et al., 1990). Figure 4-15b shows the evolution of the flat volume for the different RSLR scenarios. The No RSLR scenario shows a significant increase in the flat volume (from 28 to 73 Mm^3) due to the constant mean sea level and sediment infilling into the basin. Scenario 2 shows a relatively constant flat volume. This is due to counter balancing of new accommodation space gained due to the relatively low rise in MSL and the volume of sediment imported into the basin. In contrast, flat volume at the end of the simulation for Scenario 3 is virtually zero, indicating that tidal flats would drown under these conditions. In this case, the amount of sediment imported into the basin is larger (Figure 4-13), but still insufficient to fill the new accommodation space gained by the large rise in MSL.

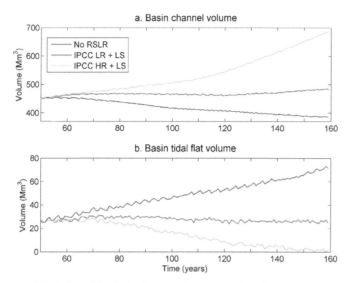

Figure 4-15 Evolution of the Basin channel volume (a), and the Tidal flat volume (b) during the 110 years simulation for the three RSLR scenarios given in Table 4-4

(c)   Flat area and height in the basin

The flat area is defined as the surface area of the tidal flats between mean low water and mean high water (Figure 4-17c). Average flat height is obtained by dividing the flat volume by the flat area. Figure 4-16 shows the evolution of the tidal flat area and flat height for the different RSLR scenarios.

The No RSLR scenario indicates an increase of about 60 Mm^2 and 0.15 m in the flat area and flat height, respectively, during the 110-year simulation (Figure 4-16a and b). This is due to the weakly flood dominant condition and constant MSL in this simulation. In Scenario 2 (low RSLR), the flat area increases marginally. This is evidence that, under these conditions, the volume of sediment imported slightly exceeds the accommodation space gained due to RSLR. In Scenario 3 (with the highest RSLR), however, both the flat area and flat height decrease significantly. This is another indication that the volume of sediment imported into the basin is inadequate to fill the accommodation space gained due to the large rise in MSL, and hence the tidal flats drown under this scenario. The highly significant reduction of the tidal flat height to almost zero (0.1 m) indicates that under these conditions, the tidal basin may degenerate into a tidal lagoon. A similar response is described in the final stage of the basin-inundation model of Oertel et al (1992). A good example of the occurrence of this phenomenon in nature is the Holocene evolution of the North Sea coastal plain described by Beets and Van der Speck (2000). The coastal plain was developed by the amalgamation of a large number of relatively small and shallow estuaries as RSLR resulted in the inundation of the valleys of the existing drainage pattern.

This process has lead to the development of tidal basin and lagoon systems depending on the sediment supply to balance the accommodation space created by the rate of RSLR.

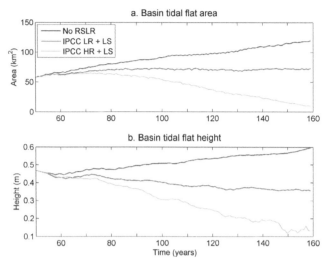

Figure 4-16 Evolution of Basin tidal flat area (a), and Tidal flat height (b) during the 110 years simulation for the three RSLR scenarios given in Table 4-4

The final predicted values (after 110 years of simulation) of the various morphological elemental characteristics discussed above are shown in Table 4-6 as net changes relative to the initial bathymetry (i.e. after the Establishment simulation). Negative values indicate a decrease compared to the initial value. In summary, Table 4-6 shows that:

- The ebb-tidal delta erodes due to RSLR. The rate of erosion is positively correlated with the rate of RSLR.

- The basin accretes due to RSLR. The rate of accretion is positively correlated with the rate of RSLR.

- Under Scenario 2 (low RSLR), the tidal flats remain relatively unchanged over the 110 years, indicating that these conditions may represent critical forcing conditions (tipping point) for the continued maintenance of the tidal flats.

- Under Scenario 3 (high RSLR), the tidal flats are drowned, possibly leading to the degeneration of the system into a lagoon.

RSLR Scenario	Net change after 110 years evolution					
	Ebb-tidal delta volume (Mm^3)	Basin volume (Mm^3)	Basin channel volume (Mm^3)	Basin tidal flats		
				Volume (Mm^3)	Area (km^2)	Height (m)
No RSLR	-23	133	-68	45	63	0.13
IPCC L + LS	-48	172	30	-2	16	-0.11
IPCC H + LS	-85	211	232	-25	-47	-0.34

Table 4-6 Morphological elemental characteristics after 110 years as net changes relative to the initial bathymetry (i.e. after the Establishment simulation) for the three RSLR scenarios given in Table 4-4

This analysis qualitatively described the RSLR effects on the inlet evolution. However, it was very limited in terms of the equilibrium state of the predicted evolutions. Therefore, these aspects are investigated comparing the RSLR scenario predicted morphologies with the empirical-equilibrium relations that describe the equilibrium state of a tidal inlet.

4.6.3   Comparing inlet evolution with empirical-equilibrium relations

Three empirical-equilibrium relations are employed; 1). Tidal amplitude to mean channel depth ratio vs. Shoal volume to channel volume ratio (Wang et al., 1999), 2). Relative flat area vs. Basin area (Eysink, 1991), and 3). Tidal prism vs. Inlet cross-sectional area (Jarret, 1978; Eysink, 1990).

Tidal amplitude to mean channel depth ratio (a/h) vs. shoal volume to channel volume ratio (V_s/V_c)

One particular characteristic of estuarine/tidal basin hydrodynamics is the distortion of tidal wave as it propagates through the shallow systems. The non-linear tidal distortion is an important parameter to tidal hydrodynamics, sediment transport and subsequently to the evolution of the shallow systems (Boon and Byrne, 1981; Aubrey and Speer, 1985). The tidal distortion inside the basin primarily depends on the frictional interaction with the channel bed and intertidal storage of the tidal flats and marshes (Friedrichs and Aubrey, 1988). The frictional interaction on the channel bed indicates by the ratio of tidal amplitude to mean channel depth (a/h) while the intertidal storage is defined by the ratio of shoal volume to channel volume (V_s/V_c). V_c is the volume of water below MSL (Figure 4-17a) and V_s is the volume of water retained on the shoals between MLW and MHW (Figure 4-17b).

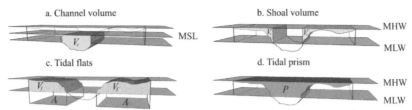

Figure 4-17 Definition of basin parameters; Channel volume (a), Shoal volume (b), Tidal flats (c) and Tidal prism (d)

Based on the findings of Dronkers (1986, 1998) and Friedrichs and Aubrey (1988), the equilibrium relationship between a/h and V_s/V_c has been extended by Wang et al (1999) to suite the Dutch estuarine systems as follows,

$$\frac{V_s}{V_c} = \frac{2\frac{a}{h} + \frac{\frac{8}{3}\left(\frac{a}{h}\right)^2}{1-\frac{a}{h}}}{\frac{3}{4}+\frac{1}{4}\frac{a}{h}}$$

4-5

where; V_s, shoal volume (m³); V_c, channel volume (m³); a, tidal amplitude (m); h, mean channel depth (m).

Figure 4-18a shows comparison of the model results in the three RSLR scenarios with the criterion in $Eq.$ 4-5. The direction of the inlet evolution is given by an arrow. The line indicates the equilibrium state while left- and right-side represent ebb- and flood-dominant systems.

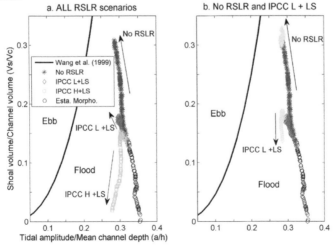

Figure 4-18 Comparison of model results with the equilibrium criterion of *Eq.* 4-5; No RSLR (*), IPCC L+LS (◊) and IPCC H+LS ( ); for 110 years (a) and 110+100 years (b)

Evolution of the flat bed (○) approaches the equilibrium relation during 50 years due to the pattern formation. The No RSLR scenario (*) predicts marginal decrease of a/h due to slight deepening of channels and increase of V_s/V_c due to sediment infilling with constant MSL. Thus, the system develops towards the equilibrium. In Scenario 2 (IPCC L + LS) (◊), the changes in both ratios are smaller implying that low RSLR steers the system by balancing the accommodation space and the sediment import into the basin. In Scenario 3 (IPCC H + LS) ( ), the highest RSLR significantly increases V_c resulting to strong decrease of V_s/V_c and the system appears to be degenerating into a lagoon.

The No RSLR and the low RSLR scenarios warrant further investigation. Therefore, these scenarios were simulated for another 100-year period assuming that the low RSLR scenario continues with the final RSLR rate of the previous 110 years (i.e. 3 mm/year). The No RSLR scenario predicts weak evolution towards the equilibrium while the low RSLR results in diverging from the equilibrium (Figure 4-18*b*). Thus, both low and high RSLR scenarios suggest drowning of the system in centennial time scale.

Relative flat area vs. basin area (A_f/A_b and A_b)

This relation describes the tidal flat evolution which indicates the highest sensitivity to the RSLR. The basin area (A_b) is the wet surface area at MHW and the flat area (A_f) is the shoal area between MLW and MHW (Figure 4-17*c*). Ratio of flat area to basin area (A_f/A_b) is defined as the relative flat area. A relation between relative flat area and basin area (*Eq.* 4-6) has been formulated by Eysink (1991) to investigate the Dutch tidal basins and estuarine systems.

$$\frac{A_f}{A_b} = 1 - 0.025 . A_b^{0.5}$$

4-6

Figure 4-19*a* shows the comparison between the model predictions and the equilibrium relation in *Eq.* 4-6. Arrow indicates the evolution direction. Similar variation in all scenarios is found at the beginning due to the marginal effect of RSLR.

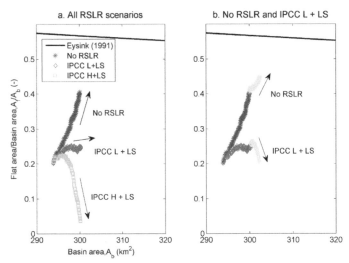

Figure 4-19 Comparison of model results and the equilibrium criterion of *Eq.* 4-6; No RSLR (*), IPCC L+LS ($\Diamond$) and IPCC H+LS ( ); for 110 years (a) and 110+100 years (b)

In No RSLR scenario (*), the flat area is increased due to sediment infilling with constant MSL. A marginal increase of the basin area ($\sim$5 km^2) is apparent in all scenarios due to widening of the inlet gorge (see Figure 3-7). In low RSLR scenario ($\Diamond$), the tidal inlet shows comparatively stable evolution due to counterbalance of the accommodation space and sediment infilling in the basin. The relative flat area significantly decreases in the scenario 3 ( ) implying that the tidal flats are drowning due to the starvation of sediment. Figure 4-19*b* shows further investigation (i.e. next 100 years) of the No RSLR and low RSLR scenarios. Weak evolution is evident in both scenarios as the morphological period increases. Similar to the previous relation, the No RSLR develops the inlet towards the equilibrium while the low RSLR tends to diverge. This also implies that the RSLR scenario predicted morphologies diverge from the equilibrium relation in centennial time scale.

Tidal prism (*P*) and inlet cross-sectional area (*A*)

P-A relationship was originally proposed by O'Brien (1931) to estimate the inlet stability. Tidal prism accounts the basin volume between the lowest low water (LLW) and the highest high water (HHW) at spring tide. However, the present analysis employed the tidal prism between MLW and MHW (Figure 4-17*d*) based on the imposed hydrodynamic boundary. Cross-sectional area is the inlet cross-section area below MSL and that varies with the RSLR scenarios. Two applications of this relation have been formulated for different environments (i.e. US Coasts by Jarret (1976) and Dutch Coast by Eysink (1990)).

The general form of the equilibrium relation is,

$$A = cP^n$$

4-7

where; A, inlet cross-sectional area below MSL (m^2); P, basin tidal prism (m^3); c and n are empirical coefficients.

Coefficient	Jarret (1976)	Eysink (1990)
c	3.797×10^{-5}	7.000×10^{-5}
n	1.03	1.00

Table 4-7 Coefficients of equilibrium relations of P and A in metric units

Figure 4-20 shows the predicted morphologies of the RSLR scenarios in comparison to the equilibrium relations in Table 4-7.

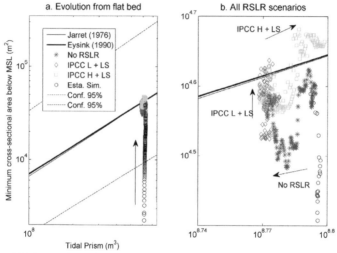

Figure 4-20 Comparison of model results and the equilibrium criterion of *Eq.* 4-7 (for both Jarret (1976) and Eysink (1990) definitions). The 95% confidence intervals are indicated by dotted lines.

The flat tidal inlet (o) has been developed towards the equilibrium state (Figure 4-20*a*). Thereafter, the results are generally scattered around the equilibrium relations and found well within the 95% confidence limits. Figure 4-20*b* shows the evolutionary trend of each scenario. The No RSLR scenario (*) results in gradual decreasing of the tidal prism due to continuous development of tidal flats with constant MSL. The scenario 2 (low RSLR) (◊) develops the inlet toward the equilibrium. This further implies counterbalance of the accommodation space and sediment infilling. In the scenario 3 (high RSLR) ( ), the model predictions overshoot the equilibrium state and the system diverges from the equilibrium relations. Such evolution is evidence of low sediment import and high increase in MSL.

Among the three equilibrium relations undertaken in this analysis, only the P-A relationship indicates equilibrium evolution of the inlet even with the non-zero RSLR scenarios. This discrepancy is most likely due to the fact that the previous two equilibrium relations explicitly account the tidal flat evolution which is highly sensitive to the RSLR.

4.7   Conclusions

Potential physical impacts of relative sea level rise (RSLR) on large inlet/basin systems with extensive tidal flats have been qualitatively investigated using the state-of-the-art Delft3D numerical model and the Realistic analogue modelling philosophy. The Ameland inlet, a typical large inlet/basin system located in the Dutch Wadden Sea was selected as a case study. Three RSLR scenarios were employed under each classification of the time-invariant RSLR ((1) *No RSLR*, (2) *5 mm/year* and (3) *10 mm/year*) and time-variant RSLR ((1) *No RSLR*, (2) *IPCC lower projection (0.2 m sea level rise by 2100 compared to 1990) and land subsidence (IPCC L + LS)* and (3) *IPCC higher projection (0.7 m sea level rise by 2100 compared to 1990) and land subsidence (IPCC H + LS)*). The model simulations spanned in decadal time scale from present-day conditions.

Model results indicate, the existing flood dominance of the system will increase with increasing the rate of RSLR. As a result, the ebb-tidal delta erodes while the basin accretes. The erosion/accretion rates are positively correlated with the rate of RSLR. Under the No RSLR condition, the tidal flats continued to develop. The time-invariant RSLR analysis suggests, the RSLR of 5 mm/year results in more or less stable evolution while 10 mm/year indicates partially drowning of the tidal flats. Under the time-variant analysis, the tidal flats eventually drowned in the high RSLR scenario, implying that the system may degenerate into a tidal lagoon. The low RSLR scenario shows similar evolution as in the case of 5 mm/year of the time-invariant RSLR implying that this may be the critical RSLR condition for the maintenance of the system. This is in contrast with the previous finding (Van Goor et al., 2003) that the tidal flats can keep up with much higher RSLR rates (up to 10.5 mm/year). This difference can be seen as an indication of the uncertainties in both approaches and should serve as a strong motivation to reduce these uncertainties via further research into the physical mechanisms governing tidal inlet response to RSLR.

As the Eustatic SLR is likely to be greater than the apparently critical rise of 0.2 m (by 2100 compared to 1990), these results indicate that the extensive tidal flats in these systems may slowly diminish, resulting in major environmental and socio-economic impacts. If the Eustatic SLR over the next century follows the higher end of IPCC projections, as has been the case over the last 15 years (Rahmstorf et al, 2007), then the tidal flats may entirely disappear converting these systems into tidal lagoons which may have such massive socio-economic impacts that the continued existence of some local communities may become untenable in the long-term. More research focusing on the quantification of the physical and socio-economic impacts of RSLR on these systems is therefore urgently needed. The lack of such quantitative information will severely hamper the development of effective and timely adaptation strategies that will enable at least the partial preservation of bio-diversity and the continued existence of local communities in these regions.

Chapter 5

Inlet effect on adjacent coastlines

Part of the material of this chapter is based on,
Dissanayake, D.M.P.K., Roelvink, J.A., Ranasinghe, R., 2009. Process-based approach on tidal inlet
evolution – Part II, Proc. International Conference in Ocean Engineering, Chennai, India. CD Rom.

5.1   Introduction

Formation of a tidal inlet due to natural events or human activities often leads to long-term and large-scale coastline changes as tidal inlets interrupt the existing alongshore sediment transport pattern. The nearshore wave climate which provides energy to mobilise sediment is in turn altered as a result of the evolving beach-inlet system. These processes might finally lead to significant impacts on the coastline adjacent to inlets resulting in socio-economic and environmental hazards. Thus, inlet effects on adjacent coastlines are important in the context of sustainable coastal management planning.

Sediment bypassing at tidal inlets is one main process that governs coastline morphology. FitzGerald et al (1978) described three main categories of sediment bypassing viz. migrating inlets, breaching ebb-tidal delta channels and stable inlets. Coastline changes associated with deep inlet gorges occur due to wave generated sediment transport (FitzGerald, 1988). The degree of inlet effects depends on the tidal range, resulting in specific morphological features (i.e. micro-tidal environment: long and relatively narrow barrier islands (Texas, eastern Florida), meso-tidal environment: short and stubby islands (Georgia, Wadden Sea)). The Ameland inlet which is the focus of this study, is located in a meso-tidal environment and consists of a deep inlet gorge (see *section 2.4*). Thus, sediment bypassing can be expected due to breaching of the ebb-tidal delta channels in combination with wave effects.

Empirical formulas (i.e. based on littoral drift and inlet hydrodynamics) are used to classify tidal inlets and in turn imply the coastline effects (Brunn and Gerritsen, 1960; Hubbard, 1976; FitzGerald et al., 1978; FitzGerald, 1988; Oertel, 1988; Bruun, 1986). These studies suggested that the inlet effect on the downdrift coastline would be greater at mixed-energy tidal inlets. Work and Dean (1990) used an analytical even/odd method to evaluate the inlet effect on the adjacent coastlines of the Florida Coast. A similar approach can be found in Fenster and Dolan (1996) who analysed the existing inlet effects on the US mid-Atlantic Coast. Castelle et al (2007) applied a numerical model and used aerial photographs to investigate the inlet effect at the Gold Coast, Australia from 1973 to 2005. Galgano (2009a,b) analysed the spatial and temporal evolution of the downdrift coastline of stabilised tidal inlets based on long-term coastline change data of the US Coast (e.g. Long Island, South Carolina, Delaware, Florida).

To date, however, the inlet effects on barrier islands of the Dutch Wadden Sea have not been fully investigated. Hence, the present analysis was undertaken with the aim of gaining qualitative insights into the inlet effects and related physical processes of the Ameland inlet. The study adopts the realistic analogue modeling philosophy in which the long term morphological evolution of a schematised inlet that closely resembles the study site is examined in detail.

5.2   Wave modules

5.2.1   General

Wave generated alongshore currents govern the littoral drift at tidal inlets. These currents are developed when waves shoal and break in shallow water. Wave effects on the sediment transport and in turn on the morphological changes are considered via online coupling of the WAVE and FLOW modules of the Delft3D modelling system (Figure 5-1) of which the wave generated forces (F_x and F_y) have been implemented in the momentum equations,

$$\frac{\partial \bar{u}}{\partial t} + \bar{u}\frac{\partial \bar{u}}{\partial x} + \bar{v}\frac{\partial \bar{u}}{\partial y} + g\frac{\partial \zeta}{\partial x} + c_f \frac{\bar{u}\left|\sqrt{\bar{u}^2 + \bar{v}^2}\right|}{h} - v\left(\frac{\partial^2 \bar{u}}{\partial x^2} + \frac{\partial^2 \bar{u}}{\partial y^2}\right) - f_{cor}v - \frac{F_x}{\rho h} = 0 \qquad \textbf{5-1}$$

$$\frac{\partial \bar{v}}{\partial t} + \bar{v}\frac{\partial \bar{v}}{\partial x} + \bar{u}\frac{\partial \bar{v}}{\partial y} + g\frac{\partial \zeta}{\partial x} + c_f \frac{\bar{v}\left|\sqrt{\bar{u}^2 + \bar{v}^2}\right|}{h} - v\left(\frac{\partial^2 \bar{v}}{\partial x^2} + \frac{\partial^2 \bar{v}}{\partial y^2}\right) - f_{cor}u - \frac{F_y}{\rho h} = 0 \qquad \textbf{5-2}$$

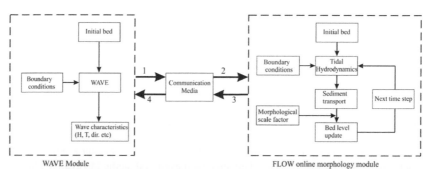

WAVE Module   FLOW online morphology module

Figure 5-1 Schematised diagram of online coupling of WAVE and FLOW modules in Delft3D with morphological updating

Initially, the wave characteristics (i.e. RMS wave height, spectral period, wave direction etc) computed by the WAVE module are temporarily stored in a communication media (Figure 5-1). Then, the FLOW module reads these wave characteristics and includes them in the flow computation. During the flow computation, bed levels are updated at every hydrodynamic time step following the *MORFAC* approach of morphodynamic upscaling. After the flow computation, the updated bathymetry and hydrodynamics (i.e. water levels, current fields) are transferred to the WAVE module via the communication media at a user defined WAVE-FLOW coupling interval. Present analysis employed two wave modules viz. SWAN and Xbeach.

5.2.2   SWAN

The spectral wind wave model SWAN (Booij et al., 1999) has been successfully applied to simulate the wave characteristics at the Ameland inlet and Wadden Sea area (Van Vledder et al., 2008; Groeneweg et al., 2008; Holthuijsen et al., 2008). SWAN is a 3rd generation wave model which simulates the physical processes of wave generation (wind), wave propagation (shoaling, refraction and frequency shift), wave dissipation (whitecapping, depth induced breaking and bottom friction) and wave-wave interactions (quadruplets and triad). The evolution of the wave spectrum is described by the spectral action balance equation in time and space (*Eq.* 5-3).

$$\underbrace{\frac{\partial N}{\partial t}}_{i} + \underbrace{\frac{\partial c_x N}{\partial x} + \frac{\partial c_y N}{\partial y}}_{ii} + \underbrace{\frac{\partial c_\sigma N}{\partial \sigma}}_{iii} + \underbrace{\frac{\partial c_\theta N}{\partial \theta}}_{iv} = \underbrace{\frac{S}{\sigma}}_{v}$$

5-3

where,
Local rate of change of wave action density in time (i), Propagation of wave action in geographical space with velocities c_x and c_y (ii), Relative frequency due to variation in depth and currents with propagation velocity c_σ (iii), Depth-induced and current-induced refraction with propagation velocity c_θ (iv) and Source term of energy density represents by the effect of generation, dissipation and non-linear wave-wave interactions (v).

Wave action density reads as,

$$N(\sigma,\theta) = \frac{E(\sigma,\theta)}{\sigma}$$

5-4

where; E, wave energy (J/m^2); σ, wave frequency (1/s); θ, wave direction (deg.)

The present analysis used the stationary mode of SWAN version 40.51A which has been coupled with Delft3D-FLOW (default). For more details of SWAN, please refer to the SWAN User Manual.

5.2.3   Xbeach

The Xbeach module also employs the wave action balance equation to describe the evolution of the wave spectrum (Roelvink et al., 2007). However, the XBeach module considers a mean frequency in the directional space and the spectral evolution is described via limited physical processes. Time and space varying action balance equation includes the wave processes of refraction, shoaling, breaking and current refraction (*Eq.* 5-5),

$$\underbrace{\frac{\partial A}{\partial t}}_{\text{i}} + \underbrace{\frac{\partial c_x A}{\partial x} + \frac{\partial c_y A}{\partial v}}_{\text{ii}} + \underbrace{\frac{\partial c_\theta A}{\partial \theta}}_{\text{iii}} = \underbrace{-\frac{D}{\sigma}}_{\text{iv}}$$

5-5

where,
Local rate of change of wave action density in time (i), Propagation of wave action in geographical space with velocities of c_x and c_y (ii), Depth-induced and current-induced refraction (iii) and Spatial dissipation of wave energy due to breaking (iv).

Wave action density reads as,

$$A(\theta) = \frac{S_w(\theta)}{\sigma}$$

5-6

where; S_w, wave energy (J/m^2)

The XBeach module solves the roller energy equation also by considering the dissipation of wave energy from the wave action balance as the source term. The roller energy further affects the wave forcing. Therefore, the wave forces are determined by the wave induced and roller induced radiation stress tensors. This analysis used the stationary version of the wave action solver, which was coupled with the Delft3D-FLOW module.

5.3   Selection of wave model parameters

5.3.1   Model set-up

Figure 5-2*a* shows the model grid consisting of fine grids (25 m × 25 m) in the inlet gorge and coarse grids (250 m × 250 m) in the open sea and basin areas. The northern boundary was selected based on available wave buoy data (i.e. buoy AZB12, see Figure 5-2*b*) while the lateral boundaries are located sufficiently away from the inlet to minimise boundary effects at the inlet. Wave conditions measured at AZB12 were used to force the WAVE module at all open boundaries. The 2004 measured data (see De Fockert et al., 2008) was used for the bottom topography (Figure 5-2*b*).

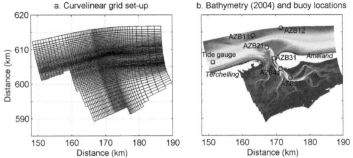

Figure 5-2 Ameland inlet curvilinear grid set-up, shown every 5th grid (a), and 2004 measured bathymetry and buoy locations (b)

Several wave-buoys have been deployed around the Ameland inlet (Figure 5-2*b*). Preliminary analysis compared the predicted and measured wave heights at these locations. The longshore wave driven processes are of utmost importance in the context of the inlet effect on adjacent coastlines. Therefore, the discussion of wave comparison is limited to the sea side buoys only (i.e. AZB11 and AZB21). The buoys AZB11 and AZB12 (i.e. forcing condition) are located seaward of the ebb-tidal delta. The buoy AZB21 is on the ebb-tidal delta. Table 5-1 shows *x* and *y* coordinates, depth of the location, buoy type and measuring frequency of these buoys. Tidal data is based on the tide gauge located north of Terschelling at a depth of ~12 m (Figure 5-2*b*).

Name	*x* (m)	*y* (m)	Depth (m)	Type of buoy	Measuring frequency (Hz)
AZB11	163100	614000	-20	DWR	1.28
AZB12	171500	616200	-20	DWR	1.28
AZB21	167300	610600	-10	DWR	1.28

Table 5-1 Buoy locations and types (DWR, Directional Wave Rider), source: Deltares report, 9S2639.A0/R0004/901483/SEP/Rott1

Wave module performance was evaluated by comparing the measured and model predicted wave heights within two storm periods, 1) between 8-10 November 2007, and 2) between 1- 3 December 2007. Figure 5-3 shows corresponding water level variations of the tidal gauge (red-line) inclusive of storm surge. The surge level (i.e. blue-line in Figure 5-3) of each storm was extracted by applying a low pass filter with 750 minutes (~ tidal period) time windows on the measured water levels.

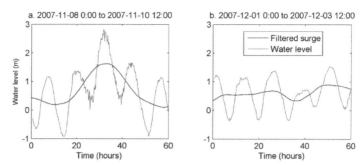

Figure 5-3 Water levels and filtered surges of the selected storm windows; from 2007-11-08 to 2007-11-10 (a) and from 2007-12-01 to 2007-12-03 (b)

5.3.2   Storm window 1: from 2007-11-08 to 2007-11-10

Figure 5-4 shows the distribution of the significant wave height around the Ameland inlet during the storm event at 14:20 on 2007-11-09 (H_s = 7.15 m, T_{m02} = 9.9 s, Dir. = 340^0 at AZB12). The dissipative effect of the ebb-tidal delta is evident on wave propagation implying that the extreme wave conditions of the North Sea hardly penetrate into the inlet. Wave energy is significantly dissipated on northern part of the ebb-tidal delta due to the rapid change of the bed topography. In contrast, more wave energy penetrates into the inlet along the western part of the ebb-tidal delta due to the orientation of the relatively deep inlet channels. Thus, strong wave driven currents can be expected at the terminal lobe and the western part of the ebb-tidal delta.

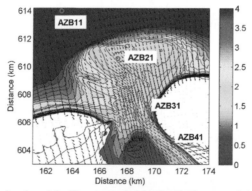

Figure 5-4 Hindcasting of significant wave height distribution around the inlet of the storm moment, 2007-11-09 14:20 (H_s = 7.15 m, T_{m02} = 9.9 s and Dir = 340 deg. nautical at AZB12)

Figure 5-5 shows the comparison between the measured and model predicted wave heights at the buoy locations in this storm window (maximum H_s~8 m). Both wave heights tend to agree at AZB11. The measured data consist of some noise which is probably due to the disturbances encountered during data acquisition. Wave heights at AZB21 are lower than that of the other buoy indicating the impact of the ebb-tidal delta on the wave propagation. However, the predicted wave heights appear to agree with the data.

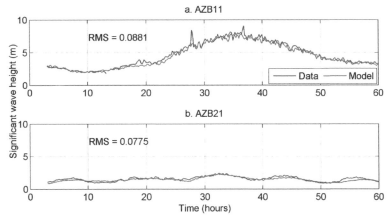

Figure 5-5 Comparison of measured data (blue) and model prediction (red) of storm window 1

Predicted wave heights were further analysed in terms of the RMS error (H_{RMS}),

$$H_{RMS} = \sqrt{\left\langle \left(H_{measured} - H_{prediction} \right)^2 \right\rangle}$$ 5-7

where; $H_{measured}$, measured wave height (m), $H_{prediction}$, model predicted wave height (m) and $\langle \rangle$ indicates mean value.

If the H_{RMS} tends to zero, the predicted wave heights are similar to the measured wave heights while larger values imply a difference. The H_{RMS} of both buoy locations are in the order of 10^{-2} implying that the model predictions (i.e. nearshore wave heights) are in good agreement with the measurements (Figure 5-5).

5.3.3   Storm window 2: from 2007-12-01 to 2007-12-03

This storm period has scattered wave heights with several peaks and the maximum (~ 4 m) is smaller than the previous storm (~ 8 m). Predicted wave heights (Figure 5-6) at AZB11 appear to have slightly underestimated compared to the data. As found earlier, wave heights are smaller on the ebb-tidal delta (AZB21). However, both predicted and measured wave heights show good agreement at this buoy.

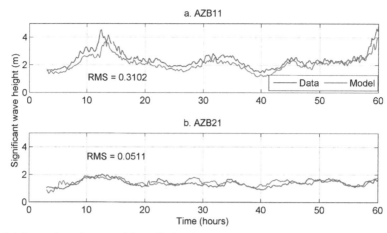

Figure 5-6 Comparison of measured data (blue) and model predictions (red) of storm window 2

5.4   Application of wave effect

5.4.1   Schematisation of wave climate

The present analysis employs a yearly wave climate at SON (north of Schiermonnikoog island; the adjacent barrier island to the east of the Ameland island) from the online wave data source of *Rijkswaterstaat* (Dutch agency of Public Works and Water Management). It should be noted that the wave climate is assumed to be stable during the morphological period of several decades (i.e. no significant climate change driven variations).

Applying a schematised wave climate is unavoidable because present computational costs do not allow multiple long term brute force simulations. Thus, different input reduction schemes are adopted to decrease the number of wave conditions. Initially, the wave conditions with probability less than 0.01% were excluded due to their rare occurrence. The remaining wave conditions were clustered into several directional classes (i.e. 0-20, 20-40...) (Figure 5-7).

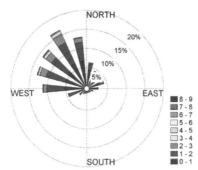

Figure 5-7 Online generated yearly wave climate at the station of SON (north of Schiermonnikoog island), Source: www.golfklimaat.nl

Secondly, the wave conditions were further screened based on the wave directions (i.e. nautical) and the orientation of the model domain. The schematised inlet is mainly exposed to waves approaching from 0 – 90 and 270 – 360 deg. directional sectors. Thus, the wave conditions from 90 – 180 and 180 – 270 deg. were also excluded. These two filtrations resulted in 49 wave conditions.

Finally, the OPTI routine (per. com. Deltares), which is based on morphological changes, was applied to select the most dominant representative wave conditions (for morphological evolution). The OPTI routine reduces the number of wave conditions based on short-term (e.g. one day) morphological response to each wave condition. Initially, the routine estimates the overall averaged morphological change due to the complete wave climate (49 conditions in this case). Then, at each iteration step, the wave condition which has the lowest contribution to the overall morphological change is eliminated. The iteration continues until one wave condition is left. The routine also determines how well the morphological change is represented by the reduced number of conditions in terms of statistical parameters (i.e. bias, RMS, R^2, standard deviation). Thus, a relatively few representative wave conditions can be selected based on these statistical parameters.

Figure 5-8 shows variation of the R^2 after removing a single wave condition at each iteration step. The last wave condition has a R^2 value of about 0.986 while it is about 0.997 for the last three wave conditions combined. The difference between these two conditions (i.e. one- and three-wave) is significant compared to that of the three-wave condition and other conditions (i.e. more wave conditions). Thus, the three-wave condition is hereon employed for the analysis.

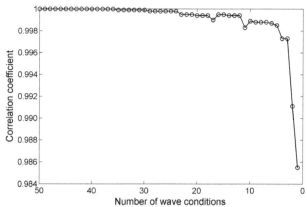

Figure 5-8 Predicted R^2 of OPTI after removing a single wave condition at each iteration

Table 5-2 shows the selected three wave conditions and their weight factors. The weight factor depends on the contribution to the overall bed level changes during the one-day simulations. The northeasterly wave with the lowest wave height (0.59 m) (W2) has the highest weight factor (0.58) and northwesterly waves with larger wave heights (i.e. 1.46 m and 3.42 m respectively) (W1 and W3) have smaller weight factors. This implies that the weight factor does not necessarily represent the dominant wave direction of the wave climate (i.e. from northwestern, see Figure 5-7).

No	Weight factor	H$_s$ (m)	T$_p$ (s)	Direction (deg. nau.)	Wind speed (m/s)	Wind direction (deg. nau.)
W1	0.33	1.46	6.3	330	5.5	307
W2	0.58	0.59	4.5	50	5.5	95
W3	0.09	3.42	7.7	290	14.2	269

Table 5-2 Characteristics of schematised three wave conditions from yearly wave climate

Preliminary morphological simulations including the wave effect demanded a large computational time. Thus, further optimisation of the model set-up is required before attempting to simulate decadal morphodynamic evolutions. Therefore, the grid resolution was decreased by a factor of 2 in both x and y directions. This approach decreased the computational time by a factor of 4 while maintaining the major morphological features of the inlet.

5.4.2   Comparison of wave modules

The SWAN grid (coupled with Delft3D) defines the x-axis alongshore and offshore directed y-axis. The Xbeach grid uses alongshore directed x-axis and onshore directed y-axis enclosing a rectangular space of the entire model area. Thus, different grid set-ups are required (Figure 5-9).

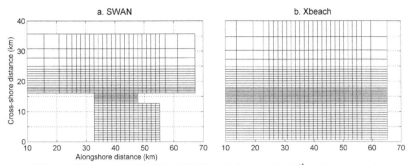

Figure 5-9 Wave grid set-up, (a) SWAN and (b) Xbeach (every other 3rd grids are shown)

The selected 3 wave conditions (W1, W2 and W3 in Table 5-2) were separately applied along the open boundaries of the model domains. The wind effect was excluded because wind-related processes are not yet implemented in the Xbeach module. Thus, the main difference between these two modules is that the SWAN module estimates wave height (H) based on directional and frequency domain ($H(f,\theta)$) while the Xbeach module uses a mean frequency in directional domain only ($H(\theta)$).

Figure 5-10 shows cross-shore propagation of the significant wave height at the alongshore location, x=25 km (see Figure 5-9). There is good agreement between the two models farther offshore while some differences can be seen in the nearshore. The largest difference is associated with the highest wave height (Figure 5-10c). The Xbeach module results in strong shoaling effect in the breaker zone. The SWAN module appears to underestimate the wave heights under depth-limited conditions as discussed in Groeneweg et al (2008). Thus, the Xbeach module tends to transfer more wave energy to the nearshore area. This would result in stronger wave driven alongshore currents and in turn alongshore transport.

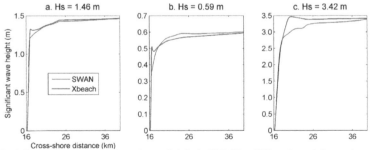

Figure 5-10 Cross-shore propagation of wave height in SWAN and Xbeach modules at alongshore distance, x=25 km, (a) H$_s$ = 1.46 m, (b) H$_s$ = 0.59 m and (c) H$_s$ = 3.42 m

Additionally, application of the Xbeach module significantly decreased the computational time (i.e. ~60% less than the SWAN module) due to the reduced number of physical processes. Thus, the Xbeach module is hereon employed to investigate the decadal evolution of the schematised inlet and the effects on adjacent coastlines.

5.5   Establishment simulation

The schematised inlet consists of a flat basin and an uniform cross-shore profile while loosely resembles the morphometry of the Ameland inlet (see *section 3.4.3*). Thus, an initial imbalance between initial model bed and hydrodynamic forcing resulting in large and unrealistic morphological changes at the beginning of the simulation is unavoidable. Therefore, as in *Chapter 4*, a morphological establishment simulation was undertaken to develop an established morphology that is more or less in equilibrium with the tidal and wave boundary forcings.

5.5.1   Selection of MORFAC

Schematisation of the wave climate resulted in three wave conditions with different weight factors (Table 5-2). Therefore, each wave condition, and their relative weights must be correctly represented in the morphological simulation. This can be achieved by implementing a variable *MORFAC* approach. In this approach, each wave condition is simulated for the duration of a fixed hydrodynamic period. A *MORFAC* specific to each wave condition is applied such that the morphological duration associated with each wave condition represents the weight factor.

The *MORFAC* required is computed by,

$$MORFAC_i = \frac{T_{Morpho}}{T_{Hydro}} \times w_i \qquad\qquad \textbf{5-8}$$

where, $i = 1,2,3$ index of the wave condition (see Table 5-2); T_{Morpho}, morphological period; T_{Hydro}, fixed hydrodynamic period; w, weight factor of the wave conditions.

Seasonal changes are neglected in this analysis due to the schematised modelling approach. Therefore, a one year morphological period (T_{Morpho}) was selected to implement the variable *MORFAC*s. By applying an iterative technique, the fixed hydrodynamic period (T_{Hydro}) was estimated as 745×5 minutes (i.e. 5 tidal cycles), and as such it is ensured that all computed *MORFAC* values are below 100 noting that the optimum bed evolution was found with the *MORFAC* of 100 under the tidal boundary forcings only (see *section 3.6.1*, see also Ranasinghe et al., 2011).

The simulation consists of hydrodynamic adaptation periods in order to change the *MORFAC* between two wave conditions and thus the same approximate suspended sediment concentrations exist with the new and old *MORFAC* values. Otherwise, a significant difference of the sediment mass will occur between 2 consecutive wave conditions (see Lesser, 2009). Accordingly, the one year morphological period is about 10 days of hydrodynamic period (i.e. 745×5×3 (wave conditions) + hydrodynamic adaptation periods). Then, this hydrodynamic period is repetitively applied to investigate the decadal evolution of the schematised inlet. Table 5-3 shows the selected *MORFAC*s and the corresponding morphological periods.

Wave condition	*MORFAC*	Morphological period (~months)
W1	46.80	04
W2	81.43	07
W3	13.29	01

Table 5-3 Selected *MORFAC* of each wave condition based on one-year morphological period

5.5.2 Effect of wave chronology

The schematised wave climate consists of three wave conditions (Table 5-2) resulting in six possible wave chronologies (i.e. sequence of application). Only three wave chronologies are investigated herein in order to show their relative importance on bed evolution (Table 5-4).

Wave chronology	Wave conditions
WC1	W1, W2, W3
WC2	W2, W3, W1
WC3	W3, W1, W2

Table 5-4 Selected three wave chronologies to investigate their relative importance

These wave chronologies were specified to simulate 50-year bed evolutions starting from the initial flat bathymetry (see Figure 3-7). The tidal boundary condition was applied as discussed under *section 3.4.4*.

Figure 5-11 shows the seaward cumulative transport through the inlet gorge. All models predict strong seaward transport because of the initial flat bed (see *section 3.4.3*). The rate of transport decreases as channels and shoals are developed and the inlet gradually widens. The WC2 condition (i.e. initial wave condition from easterly direction) shows slightly higher transport up to about 20 years. Thereafter, different transport patterns are found due to the evolution of the flat bed. At the end of the simulation, the wave chronologies with northwesterly initial wave condition (WC1 and WC3) estimate almost similar transport while WC2 results in comparatively lower transport.

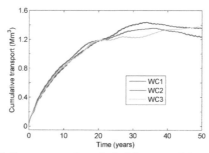

Figure 5-11 Seaward cumulative transport through the inlet gorge of the wave chronologies

The 3 model simulations above predict different bed evolutions (Figure 5-12). Thus, the wave chronology affects on the inlet evolution.

Figure 5-12a and c (corresponding to WC1 and WC3 respectively) are more comparable with respect to basin channel pattern, main inlet channel and ebb-tidal delta configuration. These bed evolutions also tend to qualitatively agree better with the Ameland inlet. However, it is difficult to discern the difference visually. Thus, statistical parameters are estimated to quantify the effect of the wave chronologies and determine the best modelled representation of the Ameland inlet.

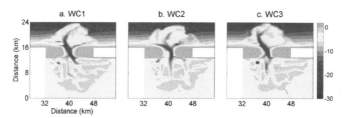

Figure 5-12 50-year bed evolution starting from the flat bed of WC1 (a), WC2 (b) and WC3 (c)

Three statistical parameters are used viz. *RMS*, R^2 and *BSS*. The *RMS* value interprets the relative bed evolution with respect to the initial flat bed while the other two parameters compare the bed level changes with the Ameland inlet.

RMS reads as,

$$z_{RMS} = \sqrt{\left\langle \left(z_i - z_{Flat}\right)^2 \right\rangle}$$

5-9

where; z_{Flat}, initial flat bed; z_i, predicted bed; $i = 1,2,3$ wave chronology index and $\langle\rangle$ indicates mean value.

If z_{RMS} tends to zero, the developed bathymetry is similar to the initial flat bed while larger z_{RMS} values imply significant evolution. Figure 5-13 shows the time evolution of z_{RMS} for the various inlet elements. The evolution of the flat bed under tidal forcing only is also included (black dashed-line) to contrast the wave effect. Initially, strong evolution occurs and the strongest/weakest evolutions are found in the inlet/basin area respectively (see y scale). This was qualitatively evident in the bed evolutions shown in Figure 5-12. However, there is no significant quantitative difference among the 3 wave chronologies. This is probably due to the fact that the wave chronology effect is overruled by the initial flat bed effect. Compared to the evolution predicted by these wave and tide forced simulations, the evolution predicted by the simulation with tidal forcing only is significantly different in the ebb-tidal delta and inlet gorge.

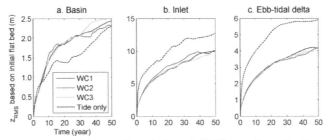

Figure 5-13 Evolution of z_{RMS} with wave chronologies and tidal force only starting from the flat bed

The R^2 and BSS values were estimated based on the control areas of the predicted beds and the Ameland inlet (see *section 4.3.2*).

R^2 value is defined as,

$$R^2 = 1 - \frac{\sum (z_{measured} - z_i)^2}{\sum (z_{measured} - \langle z_{measured} \rangle)^2}$$

5-10

where; $z_{measured}$, measured bed of the Ameland inlet; z_i, predicted bed of the wave chronologies; $i = 1,2,3$ and $\langle \rangle$ indicates mean value.

If the predicted bed is similar to the measured bed (see Figure 3-16), numerator tends to zero implying higher R^2 value. Numerator increases when the predicted bed strongly differs to the measured bed leading to lower R^2 values.

Figure 5-14a shows the temporal evolution of R^2 for the 3 wave chronologies. Initially, all models result in negative R^2 values because of the flat bed. However, the R^2 becomes positive after a few years due to the deepening of the inlet gorge. Later (after about 10 years), the R^2 slightly changes based on the wave chronology. Finally, the R^2 of WC2 decreases to about zero while the other two cases show comparatively higher values (over 0.2). Such variation can be expected based on the basin channel pattern, main inlet channel and ebb-tidal delta (see Figure 5-12) of the control area. The predicted bed of WC1 resulted in the highest R^2 implying the best representation to the Ameland inlet.

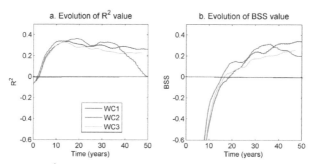

Figure 5-14 Evolution of R^2 (a) and BSS (b) values with different wave chronologies of Table 5-4

BSS value reads as,

$$BSS = 1 - \frac{\left\langle (z_{measured} - z_i)^2 \right\rangle}{\left\langle (z_{Flat} - z_i)^2 \right\rangle}$$

5-11

where, i = 1, 2, 3, wave chronology index

The *BSS* value of 1 implies an excellent comparison with the measured bed (see Figure 3-16). If the predicted bed is similar to the flat bed, denominator tends to zero leading to much lower negative values. If the predicted bed approaches the measured bed, numerator becomes lower implying the *BSS* values of nearly 1.

In contrast to the R^2, the *BSS* needs several years to reach positive values (Figure 5-14*b*). This is probably due to the fact that the R^2 directly compares the predicted and measured beds while the *BSS* compares predictions with reference to the initial flat bed. Thus, higher negative values are initially expected as the predicted bed is very similar to the flat bed. With WC1, the final *BSS* is greater than 0.3. WC2 shows a decreasing *BSS* in the final stages similar to R^2. Both parameters show the highest value with WC1 implying the best representation of the Ameland inlet. Therefore, WC1 is employed to investigate inlet effects on adjacent coastlines. The final predicted bed with WC1 (after a 50 years simulation) is hereon referred to as 'initial bathymetry'.

5.6   Inlet effects on coastlines

5.6.1   Sediment bypassing mechanism

Sediment bypassing is the key mechanism governing inlet effects on adjacent coastlines. This phenomenon was investigated by analysing the predicted bed evolutions from 50 to 100 years with WC1 (Table 5-4). Results showed a high Ω/M_{tot} ratio (i.e. $\Omega \sim 580$ Mm^3 and $M_{tot} \sim 1$ Mm^3/year) implying that tidal flow bypassing is the dominant transport mechanism (see *section 2.1*).

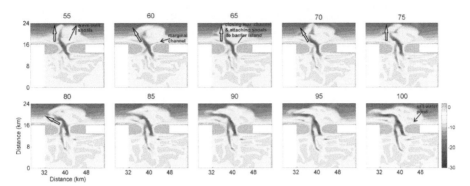

Figure 5-15 Evolution of schematised flat bed from 55 to 100 years with WC1 (Table 5-2); arrow indicates orientation of the main inlet channel on the ebb-tidal delta; location of wave built shoals, marginal channel and salt water pond are also shown.

At the start of the simulation (year 50), the main inlet channel has a northwesterly orientation and the ebb-tidal delta shows more shoal areas towards the northern end (see Figure 5-12a). This indicates seaward transport in the main inlet channel. From 50 to 80 years, the main inlet channel on the ebb-tidal delta shows a cyclic behaviour of about 10 years (i.e. changing between northwestern and northward orientation, refer to arrow on Figure 5-15). It should be noted that this cyclic evolution was not found with tidal forcings only (see *section 4.5.1*). Therefore, it is mainly governed by wave driven processes, especially alongshore transport. The shoal areas on the ebb-tidal delta tend to accumulate and move towards the eastern barrier island thus closing the marginal channel and completing the sediment bypassing process. During the cyclic evolution, the basin channel pattern develops to the east while the main basin channel approaches the eastern boundary. Thus, further evolution of this channel is hindered resulting in the pronouncement of a secondary channel to which basin flow concentrates (see year 80). Thereafter, the cyclic evolution disappears and sediment bypassing occurs along the terminal lobe developing more shoals on the ebb-tidal delta (see year 90). Finally, a two-channel system is established in the inlet and the shoal areas are connected to the eastern coastline (see year 100). These shore-connected shoals are defined as the attachment bars (Kraus, 2000). In case of incomplete attachment, these shoal areas form small salt water ponds (FitzGerald et al., 1978). Such an area appears to have developed at the north of the eastern coastline in the year 100 bathymetry. The Ameland inlet also shows a salt water pond at the eastern barrier island (Cheung et al., 2007).

The statistical parameters were further investigated during this 2nd 50 years simulation (Figure 5-16). The R^2 value remains more or less constant while the *BSS* increases from 0.34 to 0.43 (from 50 to 100 years) implying that the predicted 100-year bed more resembles the present day Ameland inlet relatively well.

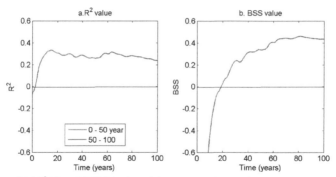

Figure 5-16 R^2 (a) and BSS (b) value of the evolution from 0-50 (red) and 50-100 (blue) years

5.6.2   Coastline evolution and relative sea level rise

Here, coastline evolution is estimated to determine the inlet effects relative to the initial bathymetry (see Figure 5-12*a*) under two RSLR scenarios (Table 5-5).

No	RSLR scenario	Description
S1	No RSLR	Mean sea level (MSL) is constant
S2	10 mm/year	Constant rate of sea level rise

Table 5-5 RSLR scenarios to investigate the inlet effects on adjacent coastlines

Figure 5-17 shows predicted bed evolutions after 50 years starting from the initial bathymetry. The S2 scenario indicates higher ebb-tidal delta erosion and basin infilling compared to the S1 scenario. Under the S2 scenario, MSL gradually rises from 0 to 0.5 m and thus wave breaking on the terminal lobe and the nearshore area move closer to the coastline as time progresses. Therefore, wave driven sediment transport gradually accelerates compared to the S1 as evidenced by the higher ebb-tidal delta erosion and basin infilling in the S2 scenario.

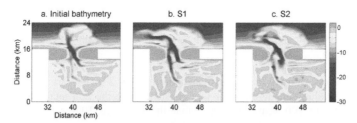

Figure 5-17 Initial bathymetry (a) and predicted bed evolutions with RSLR, S1 (b) and S2 (c)

Two indicators (i.e. cross-shore volume change and coastline position) are employed to quantify the coastline evolution on the adjacent barrier islands due to RSLR and inlet effects. The cross-shore volume change is calculated with respect to the no-inlet bathymetry (see *section 2.1*). The coastline position is estimated based on the MSL (i.e. depending on the RSLR) and foreshore slope (~1:100 on the initial bed). It is emphasised that there are erodible grid points in the alongshore direction of the model area implying a 'dry beach' (see Figure 3-7). Furthermore, the foreshore slope changes as the evolution occurs. Therefore, both MSL and foreshore slope need to be considered to find the land/water boundary of the model area which indicates the coastline position.

Cross-shore volume change

Additional 100 years long simulations were undertaken for the two RSLR scenarios with the no-inlet bathymetry (i.e. undisturbed foreshore slope). The final morphology predicted for the no-inlet bathymetry (after 50 years and 100 years, as appropriate) was used as a benchmark to compare the cross-shore volume change associated with the inlet-bathymetry simulations. If the relative volume change tends to zero, the no-inlet bed is more or less similar to the inlet-bed implying a marginal inlet effect on the coastline. Positive and negative volume changes indicate accretion and erosion areas respectively.

Figure 5-18*a* shows the estimated relative cross-shore volume change for the initial bathymetry (i.e. 50 years evolution of the flat bed). The inlet effect on the eastern coastline is larger than on the western coastline. Therefore, the model appears to be able predict the expected larger inlet effect on the downdrift (i.e. eastern) coastline (Galgano, 2009a,b; Fenster and Dolan, 1996; Dean and Work, 1993; Mann, 1993; FitzGerald et al., 1988). Significant relative changes are found only around the inlet. To the west of the inlet, a strong relative erosion is indicated due to the main inlet channel (Figure 5-17*a*). To the east of the inlet, a relative accretion area is indicated due to the ebb-tidal delta shoals associated with the inlet-bed.

Both scenarios after 100-year evolutions show almost similar relative volume changes on the eastern coastline while different patterns are found on the western (Figure 5-18*b*). To the east of the inlet, a strong relative accretion can be seen close to the inlet, while relative erosion can be seen farther away from the inlet. Under the S1 scenario, the predicted bed has a comparatively higher sediment volume implying lower erosion on the ebb-tidal delta. To the west of the inlet, both scenarios indicate northwesterly oriented main inlet channel and shoal areas (Figure 5-17). Thus, the western coast shows an erosion area close to the inlet and an accretion area farther away from the inlet. The highest erosion is found under the S2 scenario due to the lowest shoal areas and the deepest inlet channel.

Figure 5-18*c* shows the volume change from 50 to 100 years. The S1 and S2 scenarios result in net relative volume changes of about +17 Mm^3 and -26 Mm^3 respectively. This difference is probably due to the fact that, with the S2 scenario, sediment erosion gradually increases due to the gradual increase of MSL resulting in the enhancement of wave driven sediment transport. Therefore, less sediment can be expected to remain on the ebb-tidal delta when compared to the S1 scenario. These results therefore suggest that the inlet effect on the western (updrift) barrier island will be greater with non-zero RSLR rates, while the impact of RSLR on the eastern (downdrift) barrier island is negligible.

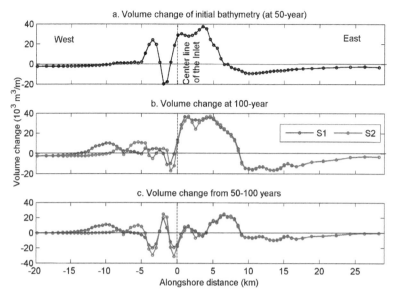

Figure 5-18 Cross-shore volume change compared to the no-inlet bathymetry; Volume change of the initial bathymetry (a), Volume change at 100-year (b), and Volume change from 50- to 100-year (c).

Coastline position

The coastline position is measured from a reference point that is located on the landward end of the cross-shore profile (i.e. cross-shore distance is zero). According to the cross-shore volume change (see Figure 5-18), an inlet effect is a certainty within the first 10 km on both sides of the inlet. Therefore, the present analysis considers high resolution (1 km intervals) beyond this range.

Figure 5-19 shows the estimated coastline positions of the initial bed and predicted beds for the 2 RSLR scenarios. The S1 scenario indicates a relatively weak evolution of the coastlines around the inlet (Figure 5-19*a*). To the east of the inlet, the wave driven transport develops shoals which tend to attach to the barrier island (see at 100 years of Figure 5-15). However, the coastline position shows erosion due to the existence of a salt water pond between the shoals and the barrier island. To the west of the inlet, the coastline shows erosion due to the orientation of the main inlet channel while it is more or less similar to that of the initial bed away from the inlet. In contrast, the S2 scenario indicates a discernible difference between the initial and final coastlines (Figure 5-19*b*). The coastline retreat which is shown for the entire coastline of the model domain, is about 50 m.

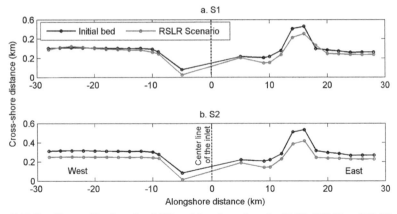

Figure 5-19 Coastline position based on MSL and foreshore slope for S1 (No RSLR) and S2 (10 mm/year) scenarios.

These results suggest that the tidal inlet has a significant influence on the adjacent barrier island coastlines, especially to the east (downdrift) of the inlet. The length of the barrier islands of the Ameland inlet is about 25 km and thus the inlet effect can be expected along the entire coastline of the islands.

5.6.3   Summary of governing physical processes

The model simulation undertaken in this study indicates a significant inlet effect on the adjacent coastlines. The downdrift coastline appears to have more affected than the updrift coastline. The physical processes that appear to given these phenomena are summarised below.

a)   *Inlet sediment bypassing*

Presence of a tidal inlet interrupts the alongshore transport pattern. The degree of the inlet interruption is a function on inlet configuration implying which results in different sediment bypassing systems (see *section 5.6.1*). The predicted morphologies show strong inlet effects on the adjacent coastlines due to the large inlet width (~4.0 km), depth (~25 m) and seaward extension of the ebb-tidal delta (~6.0 km), all of which result in a strong barrier for the alongshore transport. The long-term sediment losses of the ebb-tidal delta can be expected due to the sand transport into the basin or deposition in the progradation channels. These phenomena are accelerated by RSLR. Therefore, with RSLR, the amount of sediment available to be bypassed decreases resulting in a significant retreat of the downdrift coastline (east) while sediment deposition occurs on the updrift coastline (west).

b)   *Cyclic behaviour of the main inlet channel*

The bed evolution showed the onset of a cyclic behaviour of the main inlet channel (Figure 5-15). When the channel is oriented in the northwesterly direction, the wave generated currents penetrate into the inlet as the dominant wave direction is from the northwest. This results in higher sediment transport into the basin which leads to increased erosion of the ebb-tidal delta. In contrast, when the main channel is oriented to the north, the sediment import into the basin decreases and sediment bypassing increases. Therefore, the latter situation is favourable to maintain the alongshore transport and in turn decrease the retreat of the eastern coastline. In case of the RSLR, the sediment import into the basin is higher resulting in a significant loss of sediment from the ebb-tidal delta (see Figure 5-17).

c)   *Wave breaking on the ebb-tidal delta*

Wave breaking across the terminal lobe and the individual shoal areas (see Figure 5-4) create wave bores which retard the ebb-tidal current while augmenting the flood-tidal current. Wave built shoal areas are formed on either side of the main channel on the ebb-tidal delta due to coalescing and stacking of sediment (see Figure 5-15). These shoal areas migrate landward due to wave driven landward current. Large shoal areas are formed while moving these individual shoals (see Figure 5-15). Finally, these shoals attach to the shoreface. This mechanism determines the coastline position around the inlet and is governed by the dominant wave direction and height. The entire process is accelerated with the effect of RSLR due to waves breaking closer to the coastline.

d)   *Wave sheltering effect*

The wave sheltering effect of the ebb-tidal delta is dominant on mixed-energy coasts because of the well developed ebb-tidal delta shoals. The shallow and skewed nature of the ebb-tidal delta provides a natural breakwater for inlet coastlines. This is pronounced when waves break on the terminal lobe during the low tidal conditions. Further, the inlet ebb-currents interact with approaching waves and diminish the wave heights. These result in a decrease of wave energy on the shoreface. Therefore, the inlet effect decreases immediately downdrift of the inlet (i.e. eastern coastline).

5.7   Conclusions

Inlet effects on adjacent coastlines and the related physical processes have been investigated via numerical experiments (Delft3D) undertaken for a schematised tidal inlet which represents the geometric and hydrodynamic characteristics of the Ameland inlet. Tidal, Wave and Relative sea level rise (RSLR) effects were taken into account in the model simulations. Three representative wave conditions were selected to investigate inlet evolution based on three reduction techniques, 1) probability of occurrence (< 0.01%), 2) model orientation and wave directions, 3) relative contribution of each wave condition to morphological changes. The impact of wave chronology was also investigated.

With No RSLR, the model predicted the onset of a cyclic evolution of the main inlet channel and a large inlet effect on the downdrift coastline as expected on mixed-energy tidal inlets. The predicted final bathymetry consists of a two-channel morphology that resembles the present-day condition of the Ameland inlet. The coastline effect extends about 25 km and 15 km on the eastern and western barrier islands respectively. With an RSLR rate of 10 mm/year, the model predicted strong erosion of the ebb-tidal delta and infilling of the basin. This is most likely due to the enhanced wave driven transport processes which result from the increased MSL. The RSLR effect is considerable on the western coastline. Due to the wave sheltering effect of the ebb-tidal delta on the eastern coastline, there is no significant effect of RSLR on the eastern coastline. The main physical processes that govern the inlet effect on adjacent coastline appear to be: *sediment bypassing system, cyclic behaviour of the main inlet channel, wave breaking on the ebb-tidal delta* and *wave sheltering effect*.

Chapter 6

Process-based and Semi-empirical models on inlet evolution

Part of the material of this chapter is based on,
Dissanayake, D.M.P.K., Ranasinghe. R., Roelvink, J.A. and Wang, Z.B., 2011. Process-based and semi-empirical modelling approaches on tidal inlet evolution, Journal of Coastal Research, SI 64, Proc. of the 11th International Coastal Symposium, Szczecin, Poland, pp.1013-1071.
Dissanayake, D.M.P.K., Ranasinghe. R., Roelvink, J.A. and Wang, Z.B., (Under review). Process based vs Scale aggregated modelling of long term evolution of large tidal inlet/basin systems, Coastal Engineering.

6.1   Introduction

Different modelling approaches such as process-based models and semi-empirical models can be adopted to forecast the possible evolution of tidal inlet systems under the effect of relative sea level rise (RSLR). Process-based models are based on detailed 'process-knowledge' which describes hydrodynamic and sediment transport characteristics (and their feedbacks) using basic physical principles. These models have been successfully used to simulate coastal evolutions up to decadal time scales with tidal forcing only (Dissanayake et al., 2009c; Dastgheib et al., 2008; Van der Wegen and Roelvink, 2008) and up to inter-annual time scales with tide and wave forcing (Lesser, 2009). An alternative modelling approach is provided by semi-empirical models that are based on empirical-equilibrium assumptions which have been formulated using 'data-knowledge'. These are behaviour oriented and describe coastal evolution over time scales of centuries (Van Goor et al., 2003; Müller et al., 2007). The present study, for the first time, attempts to assess the performance of these two different types of models in predicting the evolution of large tidal inlet/basin systems to RSLR over a 50 year period.

The process-based model Delft3D and the semi-empirical model ASMITA (Aggregated Scale Morphological Interaction between a Tidal basin and the Adjacent coast) (Stive et al., 1998) are used in this study to compare and contrast the predictions obtained via the two different modelling approaches.

Chapter 3 discussed application of Delft3D for modelling decadal inlet evolution. Furthermore, Delft3D model has been successfully employed to investigate the morphological evolution in response to future sea level rise scenarios (see *Chapter 4*). These studies provide evidence of the feasibility of using process-based models for investigating long-term tidal inlet evolution, which provides more detail with respect to temporal and spatial scales of possible morphological changes. However, the model requires a large computational time due to the inherent need to resolve a large number of governing physical processes.

The ASMITA model concept is based on the equilibrium evolution of inlet elements (Van Goor et al., 2003; Kragtwijk et al., 2004). The major assumption in this model is that the *areas of inlet elements remain constant while their evolutions can be described in terms of the volume changes*. Therefore, this approach predicts the temporal evolution of the element volumes. Van Goor et al (2003) applied this model to two tidal inlets in the Dutch Wadden Sea (i.e. Eijerland and Ameland) under future sea level rise scenarios. Van Goor's (2003) results suggest that the smaller the tidal basins the better these adapt to an accelerated sea level rise. Due to the scale aggregated nature of the model, ASMITA provides very rapid solutions requiring very little computational effort.

In the present study, initially, both of the above models are applied to a highly schematised representation of the Ameland inlet in the Dutch Wadden Sea (as discussed under *Chapter 4* (tidal forcing only) and *Chapter 5* (tidal and wave forcings)), under three different RSLR scenarios (see *section 4.4.1*). Finally, both models are simulated in hindcasting mode based on the measured data of the Ameland inlet. Results are compared in terms of the inlet element evolutions as delivered by the ASMITA model.

6.2   Equilibrium assumptions of ASMITA

6.2.1   General

The ASMITA model concept was first introduced by Stive et al (1998) based on the ESTMORPH model which accounts for the interaction between tidal channels and adjacent tidal flats (Wang et al., 1998). A tidal inlet is schematised into three elements in the ASMITA model representing one state variable in each element (Figure 6-1).

Ebb-tidal delta: the total sediment (dry) volume above a fictitious sea bottom which has an undisturbed sea slope without an inlet (see Figure 2-2).

Channels: the total water (wet) volume in the basin below MLW.

Tidal-Flats: the total sediment (dry) volume in the basin between MLW and MHW.

The contribution of the adjacent coastlines to the inlet evolution is considered through '*Outside world*'.

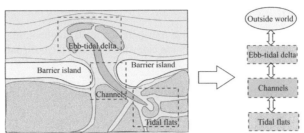

Figure 6-1 Schematisation of a tidal inlet in the ASMITA model concept (Van Goor et al., 2003)

Applications of the ASMITA model concept are extensively described by Van Goor et al (2003), and therefore are only briefly described hereon.

Evolution of a tidal inlet towards an equilibrium state depends on the sediment requirement/availability of each inlet element. This phenomenon is represented by the difference between the local equilibrium concentration (c_e) and global equilibrium concentration (c_E). The c_e is a measure of the deviation of element volume (V) from its equilibrium volume (V_e),

$$c_e = c_E \cdot \left(\frac{V}{V_e} \right)^n \qquad\qquad\qquad 6\text{-}1$$

The n power is generally taken as 2 and that is similar to the third power of the mean flow velocity, which describes sediment transport. The sign of n depends on the element volume (i.e. wet volume '+' (channels) and dry volume '-' (tidal flats and ebb-tidal delta)).

Inlet evolution has no influence on the c_E which is always in equilibrium and acts as the boundary condition of the system. Thus, if the system is in equilibrium, no sand requirement is found and the c_e of all elements is equal to the c_E. If the system is not in equilibrium, the c_e deviates from the c_E. In case, the c_e is higher than the c_E, this implies sediment surplus and erosion occurs (*negative accommodation space*). If the c_e is smaller than the c_E, sediment demand exists implying sedimentation of the element (*positive accommodation space*).

Erosion/sedimentation resulting to volume change of an element depends on the sediment availability, which is proportional to the difference between the c_e and actual concentration (c).

$$\frac{dV}{dt} = w_s . A . (c - c_e) \qquad\qquad\qquad 6\text{-}2$$

where, w_s (m/s), vertical exchange rate; A (m^2), horizontal surface area of the element.

The ASMITA model concept considers diffusive dominant transport among the elements and the sediment mass balance of an element reads as,

$$\sum \delta.\Delta c = w_s.A.(c - c_e)$$
6-3

The left-hand term reflects the diffusive transport due to all neighbouring elements, which is governed by the difference of c of these elements (Δc) and the horizontal exchange rate (δ).

Relative sea level rise (RSLR) induced volume change (V_{RSLR} (m^3)) of an element reads as,

$$V_{RSLR} = A.\frac{d\zeta}{dt}$$
6-4

$d\zeta/dt$ (m/year), RSLR.

It is emphasised that the ASMITA model considers time-invariant RSLR only. The evolution of an inlet element due to the RSLR reads as,

$$\frac{dV}{dt} = w_s.A.(c_e - c) + /- A.\frac{d\zeta}{dt}$$
6-5

Different versions of the ASMITA model have been formulated based on the number of elements viz. *one-element* (channels and outside world), *two-element* (channels, tidal flats and outside world) and *three-element* (channels, tidal flats, ebb-tidal delta and outside world). Present analysis employs the three-element version of the ASMITA model.

6.2.2   Three-element model

The fundamental mechanism that facilitates morphological evolution in ASMITA is two-way interaction between different elements driven by diffusive sediment transport. Thus, it is inherently assumed that a long-term residual sediment exchange occurs between tidal flats and channels, channels and ebb-tidal delta, and ebb-tidal delta and adjacent coastlines. This implies that the sediment supply from the 'outside world' has to pass through ebb-tidal delta and channels before reaching tidal flats. These fundamental sediment exchange patterns are described by the model formulations given below.

The volume change of the inlet elements is described by,

$$\frac{dV_i}{dt} = w_{is}.A_i.(c_i - c_{ie}) - A_i.\frac{d\zeta}{dt}$$
6-6

where, $i = d, c, f$ indicating ebb-tidal delta, channels and tidal flats respectively

The sediment availability according to the sediment mass balance reads as,

$$\delta_{od}.(c_d - c_E) + \delta_{dc}.(c_d - c_c) = w_s.A_d.(c_{de} - c_d)$$
6-7

$$\delta_{cf}.(c_c - c_f) + \delta_{dc}.(c_c - c_d) = w_s.A_c.(c_{ce} - c_c)$$
6-8

$$\delta_{cf}.(c_f - c_c) = w_s.A_f.(c_{fe} - c_f)$$
6-9

where, index od indicates outside world and ebb-tidal delta, dc indicates ebb-tidal delta and channels, c_f indicates channel and tidal flats.

The sediment demand of the elements is given by the equilibrium concentration,

$$c_{ie} = c_E \cdot \left(\frac{V_i}{V_{ie}} \right)^n$$
6-10

The equilibrium volume of each element can be defined in terms of the long-term averaged hydrodynamic parameters (i.e. tidal range R (m), tidal prism P (m³)). The equilibrium sediment volume of tidal flats is a function of basin area (A_b (m²)) and tidal range (Eysink, 1991; Eysink and Biegel, 1992; Van Geer, 2007),

$$V_{fe} = \alpha_f . A_b . R$$
6-11

The equilibrium water volume of channels is estimated using tidal prism (Gerritsen et al., 1990; Eysink and Biegel, 1992),

$$V_{ce} = \alpha_c . P^{1.55}$$
6-12

The equilibrium sediment volume of ebb-tidal delta is also a function of tidal prism (Walton and Adams, 1976; Eysink, 1990),

$$V_{de} = \alpha_d . P^{1.23}$$
6-13

The equilibrium element coefficients (i.e. α_f, α_c and α_d) are of utmost importance to determine the element parameters accurately. The tidal flat coefficient (α_f) is estimated based on the tidal flats and basin areas.

From *Eq.* 6-11, the equilibrium flat volume (Van Geer, 2007) is,

$$V_{fe} = \alpha_{fe} \cdot \left(\frac{A_{fe}}{A_b} \right) . A_b . R$$
6-14

In which,

$$\alpha_f = \alpha_{fe} \cdot \left(\frac{A_{fe}}{A_b} \right)$$
6-15

$$\frac{A_{fe}}{A_b} = 1 - 2.5 . 10^{-5} . A_b^{0.5} \qquad \text{(Renger and Partenscky, 1974)}$$
6-16

$$\alpha_{fe} = \alpha_{f0} - 0.24 . 10^{-9} . A_b \qquad \text{(Eysink and Biegel, 1992)}$$
6-17

where, V_{fe} (m³), equilibrium flat volume; A_{fe} (m²), equilibrium flat area; α_{fe} (-), equilibrium flat coefficient; α_{f0} was selected as 0.38 following Eysink and Biegel (1992) and Van Goor et al (2001).

The other two equilibrium coefficients for channels (α_c) and ebb-tidal delta (α_d) were applied as 10.9961 (Van Goor et al., 2001) and 2921.57 (per. com. with Z.B. Wang) respectively based on the measured data of the Ameland inlet.

6.3   Sensitivity of critical sea level rise

The ASMITA model describes inlet evolution by solving the set of equations (*section 6.2.2*). Table 6-1 shows the model parameters of these equations. The sensitivity to critical sea level rise ($\text{RSLR}_{\text{limit}}$) (i.e. the maximum RSLR of a system that can reach a state of morphological equilibrium) was investigated based on measured Ameland inlet data and the model parameters.

Parameter	Unit	Ameland inlet
Ebb-tidal delta area (A_d)	m^2	75×10^6
Channel area (A_c)	m^2	102×10^6
Tidal flat area (A_f)	m^2	172×10^6
Horizontal exchange rate between: outside world and delta (δ_{od})	m^3/s	1500
delta and channels (δ_{dc})	m^3/s	1500
channels and tidal flats (δ_{cf})	m^3/s	1000
Vertical exchange rate of: ebb-tidal delta (w_{sd})	m/s	1×10^{-5}
channels (w_{sc})	m/s	5×10^{-5}
tidal flats (w_{sf})	m/s	1×10^{-4}
Global equilibrium concentration (c_E)	-	2×10^{-4}

Table 6-1 Model parameters used in the sensitivity analysis of critical sea level rise

6.3.1   The Ameland inlet

The ASMITA model predicts equilibrium inlet evolution, which changes depending on the RSLR. The inlet elements acquire a new state of dynamic equilibrium if the sediment import into the system is able to fulfil the RSLR induced sediment demand (Figure 6-2).

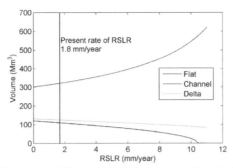

Figure 6-2 State of dynamic equilibrium of inlet elements based on the Ameland inlet data; tidal flats (blue), channels (red) and ebb-tidal delta (green) with present rate of RSLR (black)

The equilibrium element volumes of the present rate of RSLR can be found by the intersection points of the element evolution curves and the RSLR line (black). In case of the ebb-tidal delta, the equilibrium volume continuously decreases as a result of erosion due to increasing in the RSLR. The equilibrium channel volume and the rate of change increase as the rate of RSLR increases. The equilibrium tidal flat volume gradually decreases with the rate of RSLR. The tidal flats show an equilibrium state up to about 10.5 mm/year, the point which is defined as $RSLR_{limit}$. Thereafter, the tidal flats completely drown.

6.3.2   Model parameters

Element area

Figure 6-3 shows the sensitivity of $RSLR_{limit}$ to the element areas. The smaller the element area the higher the $RSLR_{limit}$. This is due to the fact that the small element area demands low sediment amount in order to reach the dynamic equilibrium state. $RSLR_{limit}$ of the tidal flat area is the most sensitive according to the inlet schematisation of the ASMITA model concept (Figure 6-1) (i.e. tidal flats are located furthest from the outside world which provides sediment into the system). Thus, it is difficult for tidal flats to satisfy the sediment demand. In contrast, the ebb-tidal delta is located adjacent to the outside world, which leads to the lowest sensitivity of $RSLR_{limit}$. The channel area shows higher sensitivity compared to the ebb-tidal delta because channel element is defined in between ebb-tidal delta and tidal flats.

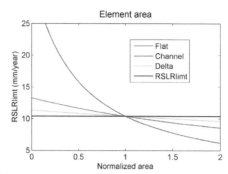

Figure 6-3 Sensitivity of RSLR$_{limit}$ to element areas; tidal flats (blue), channels (red), ebb-tidal delta (green), and RSLR$_{limit}$ of the Ameland inlet (black)

Horizontal exchange rate

Long-term diffusive sediment transport between two adjacent elements is governed by the horizontal exchange rate and the local equilibrium concentration (see *section 6.2.1*). The larger the horizontal exchange rate the higher the sediment exchange between elements in case of a difference in the local equilibrium concentrations. The horizontal exchange rate between the outside world and the ebb-tidal delta (δ_{od}) shows relatively high impact on the RSLR$_{limit}$ (Figure 6-4). Thus, if δ_{od} is larger, more sediment is imported into the system to satisfy the RSLR induced sediment demand. In contrast, the horizontal exchange rate between ebb-tidal delta and channels (δ_{dc}) and channels and tidal flats (δ_{cf}) appear to be less sensitive. RSLR$_{limit}$ of the different horizontal exchange rates can also be described in terms of the inlet schematisation of our Ameland ASMITA model (see Figure 6-1).

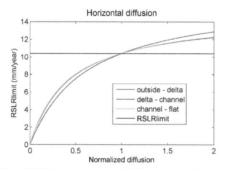

Figure 6-4 Sensitivity of RSLR$_{limit}$ to horizontal diffusion; outside to ebb-tidal delta (blue), ebb-tidal delta to channel (red), channel to tidal flat (green), and RSLR$_{limit}$ of the Ameland inlet (black)

Vertical exchange rate

Figure 6-5 shows the sensitivity of $RSLR_{limit}$ to the vertical exchange rate. This indicates relatively small impact compared to the previous parameters. However, the $RSLR_{limit}$ is extremely sensitive when the vertical exchange rate of tidal flats (w_{sf}) tends to zero. For decreasing w_{sf}, it is difficult to accrete tidal flats.

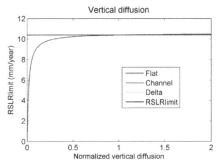

Figure 6-5 Sensitivity of $RSLR_{limit}$ to vertical diffusion; tidal flats (blue), channels (red), ebb-tidal delta (green), and $RSLR_{limit}$ of the Ameland inlet (black)

Global equilibrium concentration

The global equilibrium concentration (c_E) indicates the sediment supply into the system from the adjacent coastal stretches and offshore. Sediment availability around the inlet depends on the wave action. The wave-generated alongshore transport provides sediment close to the inlet and then tidal currents take over and transport it into the basin (see *Chapter 5*). Figure 6-6 shows the variation of $RSLR_{limit}$ by changing the c_E. The higher/smaller c_E implies strong/weak sediment supply into the system leading to strong/weak resistance to withstand RSLR.

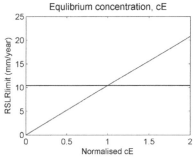

Figure 6-6 Sensitivity of $RSLR_{limit}$ to Global equilibrium concentration; Global equilibrium concentration (blue) and $RSLR_{limit}$ of the Ameland inlet (black)

6.4   Model simulations

The ASMITA model runs were undertaken for the similar morphological periods and initial conditions as in the case of the Delft3D simulations (see *Chapter 4* and *Chapter 5*). Initially, the ASMITA model was simulated for 50 and 160 years based on the schematised flat bathymetry (Table 6-2). Then, it was simulated for 50 years incorporating the time-invariant RSLR scenarios (see Table 4-3) and the established morphologies (i.e. initial bathymetry). Finally, both Delft3D and ASMITA model runs were undertaken based on the measured data of the Ameland inlet (i.e. 1970 bathymetry) to hindcast the inlet evolution from 1970 to 2004.

It should be noted that only time-invariant RSLR scenarios were investigated because preliminary runs suggested that applying several simulations to include time-variant RSLR did not provide more insight.

Model	Description			Morphological period (years)	RSLR (mm/year)
ASMITA	Initial flat bed			0 - 50	No
	Initial bathymetry	Tidal forcing only		50 - 100	No 5 10
		Tidal and wave		50 - 100	No 10
ASMITA/ Delft3D	Hindcasting from 1970 data			1970 - 2004	1.8

Table 6-2 Model simulations undertaken of the Delft3D and ASMITA model comparison (the Delft3D simulations of flat/initial bed are referred to Chapter 4/Chapter 5)

6.5   Schematised bathymetry

6.5.1   Initial flat bed

Table 6-3 shows applied input parameters of the ASMITA model based on the schematised inlet. Initially, there are no tidal flats or ebb-tidal delta on the initial flat bed. However, areas of these elements were specified according to the measured data (see Table 6-1) due to the fact that the ASMITA model assumes constant element areas throughout the simulation.

Element	Area (10^6 m^2)	Volume (10^6 m^3)	Equilibrium coefficient
Tidal flats	172	0	0.1842
Channels	102	528	11.0
Ebb-tidal delta	75	0	2921.6

Table 6-3 Element parameters and equilibrium coefficients of the ASMITA model based on the initial flat bed

The model comparison showed contrasting patterns in the element evolutions (i.e. the ASMITA model results have strong evolution in the beginning compared to the Delft3D results). This is probably due to the initial flat bed which results in strong morphological shock at the beginning of both modelling approaches. Therefore, the simulation duration of both models was extended from 50 to 160 years in order to compare the results after the initial strong evolutions.

Figure 6-7 shows the sediment transport during the morphological period of which the ASMITA model predicts strong transport pattern. In the first 50 years, the Delft3D results indicate seaward transport through the inlet gorge (Figure 6-7a). This is expected on the evolution of an initial flat bed without an ebb-tidal delta requiring sediment from the tidal basin and the adjacent coastlines. Later, the transport reaches a quasi-stable condition due to the pattern formation on the flat bed. The extended morphological period (i.e. from 50 to 160 years) shows a clearly flood-dominant transport system in the inlet. In the first 50 years, the ASMITA results show strong sediment transport from outside world to ebb-tidal delta (O-E) and channel to tidal flats (C-F) leading to develop ebb-tidal delta and tidal flats respectively (Figure 6-7b). The sediment transport from ebb-tidal delta to channel (E-C) is lower than that of O-E implying the growth of the ebb-tidal delta. The strong sediment transport pattern among the inlet elements has been gradually decreased and tends to reach stable condition at the end of this period. Finally, the sediment exchange shows that the total sediment amount receiving into the ebb-tidal delta is almost transferring into the channels. In the mean time, the transport of C-F significantly decreases and remains stable. This is probably due to the large sediment requirement of channels (i.e. whole basin area of the flat bed) before developing tidal flats. Therefore, the ASMITA model also predicts flood-dominant transport pattern of the inlet. Accordingly, both model results agree with the present-day sediment circulation pattern of the Ameland inlet.

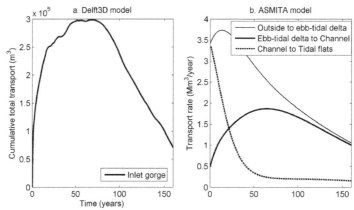

Figure 6-7 Cumulative seaward transport through the inlet gorge of the Delft3D simulation and Sediment exchange between inlet elements of the ASMITA simulation starting from the flat bathymetry; Outside world to Ebb-tidal delta (thin-line), Ebb-tidal delta to Channels (thick-line) and Channels to Tidal flats (dash-line)

Figure 6-8 shows the corresponding evolutions of the inlet element areas on the flat bed. The element areas of the Delft3D results have been gradually developed and the rate of evolution decreases along the morphological period. However, the ASMITA model assumes constant element areas throughout the model duration. The predicted element areas of the Delft3D model tend to approach their counterparts of the ASMITA model. Specifically, the ebb-tidal delta area strongly agrees compared to the other two elements because of the rapid growth of the ebb-tidal delta as witnessed with the sediment transport pattern (Figure 6-7).

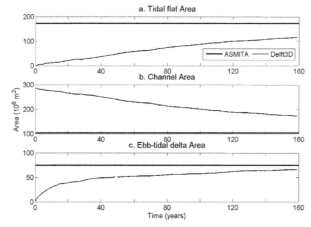

Figure 6-8 Evolution of element areas starting from the flat bed in the ASMITA and Delft3D models

Figure 6-9 shows the evolution of element volumes and tidal flat heights in both modelling approaches. The evolution suggests that both approaches develop tidal flats and ebb-tidal delta on the flat bed while decreasing the channel volume. In contrast to the first 50 years period, the evolution of flat height and element volumes of the ASMITA model widely corresponds with the Delft3D results. In the basin, the flat height is in a good agreement compared to the other element parameters. However, the flat volume initially shows some difference and that decreases during the simulation period. This contrasting behaviour of the tidal flat evolution can be described in terms of the flat areas in both models. In case of the Delft3D modelling approach, the flat area is gradually developed as the simulation progresses and so does the flat volume. Therefore, higher flat height (i.e. V_f/A_f ratio) can be expected even at the beginning of the Delft3D run. On the other hand, the ASMITA model assumes constant flat area while the volume changes during the evolution. Therefore, the flat height is governed by the temporal flat volume and constant flat area. In this context, the flat volume difference can be expected though there is a good agreement with the flat height. The channel volume shows more or less similar evolution in contrast to the tidal flat volume. Initially (up to about 50 years), the Delft3D modelling approach results in the strongest development of the ebb-tidal delta volume due to the pattern formation on the flat bed and widening of the inlet gorge. Thereafter, the ebb-tidal delta indicates a quasi-stable condition and then decrease in volume as a result of sediment import into the basin implying a flood dominant inlet system. Similar characteristics tend to have produced in the ASMITA model. The ebb-tidal delta volumes show good agreement towards the end of the simulation period.

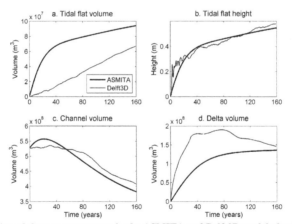

Figure 6-9 Evolution of element parameters in the ASMITA and Delft3D models for 160 years starting from the initial flat bed

Both modelling approaches suggested strong bed evolution at the beginning due to the initial flat bathymetry. Therefore, the comparison adopts the established morphologies discussed in *Chapter 4* and *Chapter 5* to investigate the responding inlet evolution under RSLR.

6.5.2   Established morphology

Established morphology is a 50 years evolution of the schematised flat tidal inlet, which is analogous to the channel/shoal pattern of the Ameland inlet. Two established morphologies were employed in this analysis (Figure 6-10). Each morphology corresponds to the boundary forcing types (i.e. tide only (see *Chapter 4*) and, tide and wave (see *Chapter 5*)).

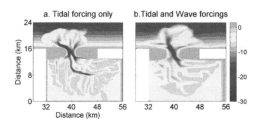

Figure 6-10 Evolution of initial flat bed for a 50-year period (Established morphology) with different boundary forcings; Tide only (a) (see Chapter 4) and Tide and wave (b) (see Chapter 5)

Tidal boundary forcing only

a.   Application of established morphology parameters as initial state

Table 6-4 shows the initial element parameters of the ASMITA model simulations based on the tidal established morphology (see Figure 6-10a).

Element	Area (10^6 m^2)	Volume (10^6 m^3)	Equilibrium coefficient
Tidal flats	59	24	0.1808
Channels	224	454	11.0
Ebb-tidal delta	50	169	2921.6

Table 6-4 Element parameters and equilibrium coefficients based on tidal established morphology

The established morphology consists of the major channel/shoal patterns of the Ameland inlet (BSS>0.3, see *Chapter 4*). However, it is emphasised that the tidal flats have not been developed as found in the 2004 measured data (i.e. measured volume ~ 120 Mm^3 and model prediction ~ 24 Mm^3). This in turn affects the channel parameters. Therefore, the established morphology has less tidal flats and more channels compared to the measured data of the Ameland inlet. In contrast, the ebb-tidal delta volume is about 169 Mm^3 of the established morphology, which is, to a first order, comparable, to the measured ebb-tidal delta volume of about 130 Mm^3.

Results of the ASMITA model showed higher sediment transport from E-C than that of O-E implying erosion of the ebb-tidal delta (*not shown*). This phenomenon has been accelerated as the rate of RSLR increases. Thus, the highest RSLR (10 mm/year) resulted in the highest erosion on the ebb-tidal delta. These characteristics were found with the

Delft3D model results too. At present, the Ameland inlet has a flood dominant (landward) transport system (Dronkers, 1998). Therefore, both model results further agree with the existing situation. Both ebb-tidal delta and tidal flat areas decrease as the rate of RSLR increases in the Delft3D model. However, the ASMITA model assumes constant element areas irrespective to the rate of RSLR.

Figure 6-11 shows volume change of the inlet elements and evolution of the flat height with respect to different RSLR scenarios in both modelling approaches. Inlet evolution in both models is very sensitive to the RSLR scenarios. The higher the rate of RSLR, the lower the tidal flat and the ebb-tidal delta growth. The channel volume increases as the rate of RSLR increases. The tidal flat evolution shows significant difference in the two approaches. This is probably due to less tidal flats and more channels at the beginning of the simulations representing a state that is quite away from the equilibrium. Therefore, a strong sediment transport from channels to tidal flats is apparent leading to a strong development of tidal flats. Furthermore, the ASMITA model used smaller flat area compared to the measured data. This is probably resulting to the surprising tidal flat heights (i.e. above 1 m) in the ASMITA model. However, the Delft3D model results show gradual evolution of tidal flats corresponding to the applied rate of RSLR. Evolution of the channels and the ebb-tidal delta volumes tends to converge in both models compared to that of the tidal flats.

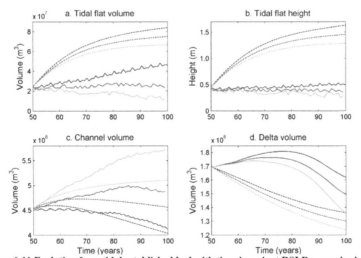

Figure 6-11 Evolution from tidal established bed with time-invariant RSLR scenarios in the ASMITA (dash lines) and Delft3D (solid lines) models; No RSLR (blue); 5 mm/year (red); 10 mm/year, (green)

These results suggest that the tidal flat evolution in the ASMITA model is very sensitive to initial element areas of the channels and tidal flats. This phenomenon was further investigated by decreasing the channel area while increasing the tidal flat area. Resulting evolution of the tidal flat height tends to improve (i.e. below 1 m). Accordingly, more realistic channel and tidal flat areas should be incorporated in the ASMITA model instead of the areas from the established morphology. Therefore, equilibrium channel area (A_{ce})

and tidal flat area (A_{fe}) (Renger and Partenscky, 1974) are hereon employed because the ASMITA model estimates equilibrium evolution of the tidal inlet.

b.   Application of equilibrium channel and tidal flat areas as initial state

The A_{fe} was estimated using the relations in *Eq.*   **6-16** and   **6-17** and then the A_{ce} can also be determined because the basin area is constant (i.e. $A_b = A_{fe} + A_{ce}$). In this analysis, all other initial parameters are similar to the previous application.

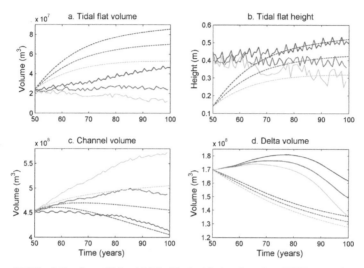

Figure 6-12 Evolution from tidal established bed with time-invariant RSLR scenarios and equilibrium channel and tidal flat areas in the ASMITA (dash lines) and Delft3D (solid lines) models; No RSLR (blue); 5 mm/year (red); 10 mm/year (green)

Applying the equilibrium element areas, the predicted tidal flat heights of the ASMITA model tend to produce good agreement with the Delft3D model results (Figure 6-12). Furthermore, tidal flat volumes appear to be lower depending on the rate of RSLR. The higher the rate of RSLR the lower the tidal flat volume. This is probably due to increased sediment demand of tidal flats under increased area and RSLR (see *Eq.*6.4). These results indicate the ASMITA model should be run applying more realistic element areas though they are assumed to be constant. The estimated equilibrium areas are nearly similar to the measured element areas of the Ameland inlet and therefore are hereon employed.

Tidal and wave boundary forcings

Due to the grid resolution and wave effect, 50-year bed evolution of the flat bed resulted in different established morphology compared to that of the tidal forcing only (see Figure 6-10b). Table 6-5 shows the initial input parameters of the ASMITA model based on the wave established morphology. The measured element areas of the Ameland inlet were employed following the previous analysis. The wave-established morphology consists of higher channel volume and lower tidal flat volume than the tidal-established morphology (Table 6-4). Furthermore, the ebb-tidal delta volume has been decreased due to the wave effects (i.e. decreasing the basin sediment supply into the ebb-delta area and advecting sediment away from the ebb-tidal delta) and the coarse grid set-up.

Element	Area (10^6 m^2)	Volume (10^6 m^3)	Equilibrium coefficient
Tidal flats	172	20	0.1842
Channels	102	590	11.0
Ebb-tidal delta	75	156	2921.6

Table 6-5 Input parameters of the ASMITA model based on the wave-established morphology

Figure 6-13 shows the resulting evolution of the inlet parameters of both modelling approaches based on the wave-established morphology. The Delft3D model under tidal and wave boundaries predicts increasing of tidal flat volume under the No RSLR scenario (Figure 6-13a). This is probably due to the fact that wave generated currents in the inlet gorge strongly decrease the seaward sediment transport while enhancing the landward transport with constant MSL. The RSLR scenario results in more or less similar flat volume as in the case of the tidal forcing only. Therefore, the model results suggest that the wave-effect has been weakened by the RSLR with respect to the tidal flat evolution. The ASMITA model predicts marginal decrease in the tidal flat evolution compared to the previous analysis, probably due to lower initial flat volume. Resulting flat volumes in both models appear to converge under the No RSLR scenario while diverging with the RSLR effect. The Delft3D results in comparison to the tidal forcing only suggest that the inclusion of wave effect increases tidal flat height (~ 0.1 m) under the No RSLR scenario while indicating more or less similar variation with the RSLR (Figure 6-13b). Initial difference of the flat height (~ 0.3 m in Figure 6-13b) in both models is expected due to the different areas of tidal flats (i.e. Delft3D ~ 40 km^2 and ASMITA ~ 172 km^2). It is emphasised that the flat area is constant in the ASMITA model while it develops in the Delft3D model. Under the RSLR, the flat heights appear to agree on the final predicted beds such that the Delft3D predicts gradual decrease (~ 0.2 m) and the ASMITA predicts marginal increase (~ 0.1 m) during the morphological period. Both model results under the No RSLR indicate that the flat height might agree if the simulation spanned for a longer period.

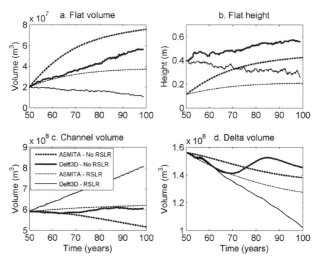

Figure 6-13 Evolution from 50-year wave-established morphology with time-invariant RSLR scenario and measured channel and tidal flat areas in the ASMITA (No RSLR – thick-dash-line; 10 mm/year – thin-dash-line) and Delft3D (No RSLR – thick-solid-line; 10 mm/year – thin-solid-line) models

Evolution of the channel volume shows higher values of the Delft3D predictions compared to that of the ASMITA (Figure 6-13c). Under the RSLR, the Delft3D model predicts strong increase (~ 200 Mm^3) while the ASMITA results show more or less stable evolution. The former implies lower sediment import into the basin compared to the increase in MSL and the latter indicates that the amount of sediment import into the channels is almost similar to the volume increase due to RSLR. Under the No RSLR, both models show similar evolution from 50 to 65 years. Thereafter, the channel volume of the Delft3D model increases and reaches stable state of which there is a relatively high agreement with the predicted volume of the ASMITA model applying the RSLR. The channel volume gradually decreases (~ 100 Mm^3) under the No RSLR case of the ASMITA model.

Resulting ebb-tidal delta volume of the Delft3D model shows similar evolution in the first decade due to lower increase of the MSL (Figure 6-13d). Under the No RSLR, the results indicate an oscillatory evolution. This can be described by the cyclic evolution of the main inlet channel and northwesterly dominant wave conditions. When the main channel is oriented to the northwesterly direction, the wave-generated currents augment the landward transport resulting strong erosion on the ebb-tidal delta. Furthermore, such orientation attenuates the seaward transport. These are evidence of the lower ebb-tidal delta volume. The inverse behaviour of these processes was found with the northward orientation of the main channel leading to increase seaward transport and thus resulting higher ebb-tidal delta volume. After inclusion of the RSLR, the cyclic behaviour of the main channel disappeared and the ebb-tidal delta continuously eroded. The ASMITA results show decrease of the ebb-tidal delta volume in both scenarios. The volume decrease of the RSLR scenario is about 20 Mm^3 higher than that of the No RSLR and further it is about 20 Mm^3 lower compared to that of the Delft3D results. However, in the first decade, the predicted volume of the ASMITA under the RSLR tends to agree with the Delft3D volume. Predicted evolution of the No RSLR indicates that both model results might converge under a longer morphological period.

6.6   Measured bathymetry

6.6.1   Comparison of initial beds

Figure 6-14 shows the 1970 measured data of the Ameland inlet (c) in comparison to initial beds of the schematised inlet (i.e. tide only (a) and, tide and wave (b) boundary forcings), which were applied in the previous analysis. It is noted that the schematised beds are based on the rectangular grid set-up (see *section 3.4.2*) while the Ameland bed is on the curvelinear grid set-up (see Figure 5-2). Both predicted beds tend to agree with the Ameland bed with respect to the main channel/shoal configuration around the inlet gorge (see *section 3.6.1* and *5.5.2*) whereas the basin has lower intertidal areas. Extent of the intertidal areas affects on the basin tidal prism. The initial beds have tidal prisms of about 576 Mm^3 (tide only) and 574 Mm^3 (tide and wave) while it is about 479 Mm^3 of the measured bed. These are evidence of the lower intertidal areas on the predicted beds. The marginal difference of the tidal prism between two forcing types is related to the grid resolution of the model domain. The higher grid resolution (i.e. tide only case) results in more channel areas leading to slightly higher tidal prism.

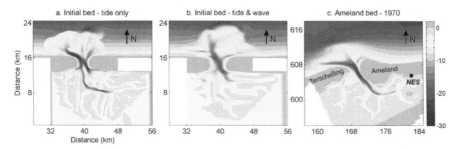

Figure 6-14 Initial beds of the schematised inlet, tide only (a), tide and wave (b) and the 1970 Ameland inlet bed (c)

Bed topography of the basin area is further analysed in terms of the area-depth hypsometry (Figure 6-15). At deep water (> 10 m depth), the tidal only case shows good agreement with the measured data due to the deeper basin channel compared to the case of both forcing types. At shallow water (i.e. intertidal areas), both predicted beds show similar characteristics and that indicates higher channel area in comparison to the data. Furthermore, the tide only case results in marginally higher basin area as found with the tidal prism. These results suggest that the intertidal areas of the predicted beds are not developed as in case of the measured data.

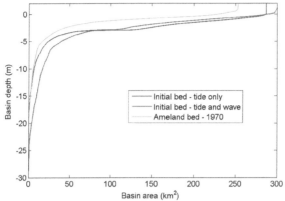

Figure 6-15 Comparison of basin hypsometry of the initial beds (established morphology) tide only (blue-line), tide and wave (red-line) and 1970 Ameland bed (green-line)

6.6.2   Hindcasting from 1970 to 2004

Measured bathymetric and water level data of the Ameland inlet area were analysed from 1970 to 2004 to investigate the bed evolution and inlet elements. There are no bathymetric data in some years and areas. In such cases, missing data were estimated using the linear interpolation based on the adjacent data sets. The water level records were used from the station, '*NES*', which is located in the basin close to the southern coast of the Ameland island (see Figure 6-14c). The evolution of the inlet elements was analysed based on the measured water levels (Table 6-6) and the transformed bathymetric data on to the model grid (see *section 3.4.2*).

Year	1970	1980	1990	2000
MLW (m)	-1.17	-1.28	-1.12	-1.18
MHW (m)	1.02	1.03	1.09	1.07

Table 6-6 Observed water levels at station *NES* (see Figure 6-14c) (source: per. com. with Deltares)

Both Delft3D and ASMITA models were simulated from the 1970 bathymetry incorporating 1.8 mm/year of RSLR (see Stive et al., 1991; Dronkers, 1998; Van Goor et al., 2003) to hindcast the inlet evolution up to 2004. Table 6-7 shows the initial input parameters and equilibrium coefficients of the ASMITA simulation. The tidal flat coefficient (α_f) was estimated applying *Eq.* **6-14** to **6-17** while the other two coefficients (i.e. α_c and α_d) are referred to *section 6.2.2*. The Delft3D model was simulated employing the tidal boundary forcing only.

Description	Element	Area $(10^6 \, m^2)$	Volume $(10^6 \, m^3)$	Equilibrium coefficient
Hindcasting from 1970 to 2004	Tidal flats	171	108	0.1850
	Channels	101	232	11.0
	Ebb-tidal delta	75	168	2921.6

Table 6-7 Initial conditions and equilibrium coefficients of the ASMITA run for hindcasting the Ameland inlet from 1970 to 2004

Resulting sediment transport rates of the ASMITA model initially indicate erosion of the ebb-tidal delta and channels implying strong growth of the tidal flats (*not shown*). Later, the transport rates gradually decrease and reach quasi-stable state. However, the ebb-tidal delta continuously erodes. The Delft3D model predicts seaward transport through the inlet gorge during the simulation. Predicted bed evolution is qualitatively compared with the initial (1970) and final (2004) beds of the measured data. The element evolution was analysed based on the model predicted bed levels and water levels.

Bed evolution

Figure 6-16 shows the predicted bed (c) of the Delft3D model in comparison to the initial (a) and final (b) beds. From 1970 to 2004, data indicate changing the inlet from one-channel system to a two-channel system which has a northward directed main inlet channel (see *section 2.4*). The main channel in the basin appears to be strong with the one-channel system of which flow is concentrated compared to the case of two-channel (see Figure 6-16*a* and *b*). Predicted bathymetry still has a one-channel system. At seaside, westward oriented inlet channel has been pronounced while weakening the northward channel. Furthermore, strong sand accumulation is apparent on the ebb-tidal delta. At basin side, more channel areas appear to have developed and the channel adjacent to the Ameland island tends to be stronger implying the direction of tidal wave propagation into the basin (i.e. from west to east alongshore tide, see *section 3.6.2.3*). The latter phenomenon is dominant on the predicted bed because the model used the tidal boundary only. For the same reason, it can be expected strong ebb-tidal delta growth also (i.e. strong seaward transport as a result of non-inclusion of the wave effect, see *section 3.6.1*).

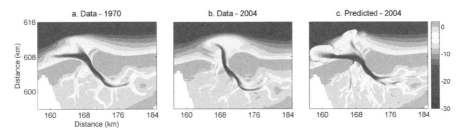

Figure 6-16 Initial (a) and final (b) beds based on data and predicted bed (c) of Delft3D

Evolution of inlet elements

Figure 6-17 shows evolution of the element volumes and tidal flat heights in both models and data. Data indicate growth of tidal flats from 1975 to 2004. Predicted flat height and volume marginally agree with the data. Initially, the ASMITA model predicts strong evolution of tidal flats (see Figure 6-17*a* and *b*). However, the flat volume tends to converge to the Delft3D results while the flat height in both models is almost similar on the final predicted beds. The Delft3D model predicts decrease of flat volume (~10 Mm³). Flat height initially decreases and then marginally increases (~0.04 m). This is probably due to the definition of the flat height (i.e. V_f/A_f) which results in higher value if the A_f strongly decreases compared to V_f. The models have overestimated the channel volume and the data indicate comparatively constant volume (Figure 6-17*c*). Both model predictions show similar trend in increasing the channel volume whereas the Delft3D results are about 10 Mm³ lower compared to the ASMITA model. Channel volume increases due to erosion (i.e. moving sediment from channel to tidal flats and ebb-tidal delta) and RSLR. The ebb-tidal delta volume shows comparatively good agreement between data and the ASMITA results (Figure 6-17*d*). Furthermore, the initial evolution (up to about 1977) indicates quite similar behaviour in all three cases. Then, the Delft3D model predicts strong increase of the ebb-tidal delta volume. The reason for such an increase is discussed by considering few snap shots during the morphological period (see 1, 2, 3, and 4 in Figure 6-17*d*).

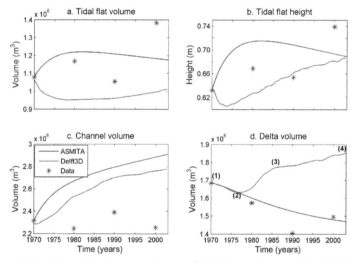

Figure 6-17 Evolution of the element volumes and tidal flat height from 1970 to 2004 in ASMITA model (blue-line), Delft3D model (red-line) and data (black-star)

Figure 6-18 shows the evolution of ebb-tidal delta at the selected time points. At 1 (1970), there are two channels on the ebb-tidal delta delivering sediment to west and north respectively. From 1970 to 1977, it appears to have deepened these channels while delivering sediment to the terminal lobe of the ebb-tidal delta (see at 2 (1977)). As such, it can be expected marginal decrease of the volume (see Figure 6-17*d*). After 1977, the northward channel shows strong sedimentation due to strong formation of shoal areas which restrict moving sediment away from the ebb-tidal delta. Thus, the volume indicates rapid growth after 1977. Formation of shoals continues while deepening the main channel. At 3 (1986), the ebb-tidal delta shows more shoal areas and deepened channel system than that of 1977. Such behaviour attributes to decrease the rate of growth as evident at 3 (1986) (see Figure 6-17d). Thereafter, bed evolution indicates that the shoals are further developed and the northward channel gradually disappears. At 4 (2004), it is evidence of further growth of the shoal areas and closing the northward channel. These characteristics lead to continuous increase of the volume as shown at 4 in Figure 6-17d.

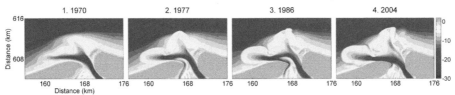

Figure 6-18 Evolution of the ebb-tidal delta at 1970 (1), 1977 (2), 1986 (3) and 2004 (4) years (see Figure 6-17d)

The ebb-tidal delta volume was estimated based on a no-inlet bathymetry which has an undisturbed coastal slope (see *section 2.1*). In case of the Ameland inlet, the no-inlet bathymetry explicitly depends on the orientation of the barrier islands because their coastlines are not aligned. Therefore, the estimated volumes might contain some uncertainties. On the other hand, it is rather certain to define a no-inlet bathymetry based on the schematised inlet of which the barrier islands are inline with the west-east direction (see Figure 6-14).

The Delft3D run was undertaken employing tidal boundary forcings only. Resulting evolutions showed unrealistic seaward expansion of the ebb delta, possibly due to the fact that wave boundary forcing was not included. Therefore, it is recommended to carry out further investigation applying both tidal and wave boundary forcings.

6.7   Conclusions

Present analysis compared long term tidal inlet evolution of the process-based model (Delft3D) and the semi-empirical model (ASMITA) based on a schematised inlet and the Ameland inlet. Evolution of the schematised inlet was investigated applying a flat bed and an initial bed which is analogous to the Ameland inlet, with three time-invariant RSLR scenarios (i.e. No RSLR, 5 mm/year and 10 mm/year) and both tidal and wave boundary forcings. The Ameland inlet evolution (from 1970 to 2004) was undertaken with 1.8 mm/year RSLR and tidal boundary forcing only.

Initial adaptation period of the ASMITA model was clearly evident after extending the simulation for a long morphological period (i.e. from 50 to 160 years). Results of both models indicated convergence of element evolutions implying that the inlet is developed towards the state of equilibrium/stable in either case. Furthermore, the predicted inlet evolutions resulted in flood dominant systems which strongly agree with the contemporary data of the Ameland inlet. Prescribing realistic element areas of the ASMITA model is prerequisite though the areas are assumed to be constant. The tidal flat evolution showed the highest sensitivity to RSLR in both modelling approaches. The initial adaptation period of the ASMITA model appears to have increased with the rate of RSLR and thus the predicted evolutions of the RSLR scenarios tend to weakly agree in the 50 years period. The Delft3D model predicted more or less stable tidal flat evolution under the 5 mm/year scenario. It is generally found that the element evolutions of both modelling approaches weakly agree after including the wave effect into the Delft3D model. In respect to the tidal flat evolution, wave effect is stronger in case of No RSLR compared to the RSLR (i.e. 10 mm/year). Thus, the tidal flat evolution with the RSLR and wave effect tends to be similar to that of the tidal forcing only.

In the hindcasting mode of the Ameland inlet (from 1970 to 2004), the predicted tidal flat evolutions of both models appear to be converging and they are lower compared to the data. On the final predicted beds, the flat height is almost similar in both modelling approaches. The channel evolution also appears to be similar in the models whereas it has been overestimated compared to the data. The ebb-tidal delta evolution of the ASMITA model tends to agree with the data while the Delft3D model predicted strong increase which is probably related to the non-inclusion of the wave effect. Therefore, it is recommended further investigation of the inlet evolution applying both tidal and wave boundary forcings.

Process-based model (Delft3D) results tend to agree with the empirical equilibrium relations of the semi-empirical model (ASMITA) to some extent after the initial adaptation periods of both models and it strongly depends on the initial conditions (i.e. bed topography   and   boundary   forcings)   and   the   RSLR   scenarios.

Chapter 7

Sand nourishment on tidal inlets

7.1   Introduction

Sand nourishment is an increasingly adopted measure for the sustainable management of coastal environments (i.e. beach and dune protection, restoration for short-term emergencies (storm induced erosion) and long-term issues (chronic erosion and RSLR effects)). Nourishments near tidal inlets are neither easy to sustain nor well-established. The application of numerical models may support the planning of nourishments. More efficient nourishment schemes can be implemented by analysing different nourishment strategies and therefore the model results support to reduce the failures and costs by assessing the environmental/economic implications (Capobianco et al., 2002).

In the European perspective, the sand nourishment is mainly applied based on three objectives, 1). improving the coastal stability, 2). improving the coastal protection and 3). increasing the beach width (Hamm et al., 2002). Specifically, nourishment application is to maintain the coastline position according to the Dutch coastal policy of 1990 (Anonymous, 1990). Thus, the sand nourishment is extensively applied on the Dutch Coast (i.e. shoreface and beach) to maintain the coastline position (~6 Mm^3/year since 1991) (Hanson et al., 2002). The shoreface nourishment acts as a reef with a lee-side effect shoreward of the nourishment area (Van Duin et al., 2004). This ultimately helps to a sustainable development of the beach area leading to net gain of sediment. Grunnet et al (2004 and 2005) used a numerical approach (Delft3D) to investigate the behaviour of the shoreface nourishment on the barrier island of Terschelling. Results of their 2DH approach suggested the same dependency on the spatial scale morphodynamics as in case of the 3D approach. Recently, sand nourishment has been investigated to compensate 'sand hunger' in the Eastern Scheldt of the Netherlands due to construction of the storm surge barriers (Camille, 2010; De Ronde et al., 2009; Escaravage et al., 2009). Camille (2010) analysed the possible morphological changes on *Galgeplaat* (i.e. an intertidal area) based on the bathymetric surveys and Argus video images and that implies positive effects of the nourishment.

To date, more efforts have investigated nourishment to mitigate the chronic erosion of coastlines and little is known on the effect of intertidal areas. RSLR results in additional sediment demand on the intertidal areas of tidal basins to follow the increased MSL (see *Chapter 4*). Thus, the additional supply of sediment into the basin has to be the result of strong erosion of adjacent coastlines and ebb-tidal delta. The efficacy of nourishing ebb-tidal delta to mitigate these RSLR induced erosive impacts on ebb-tidal delta itself and adjacent coastlines and to fulfill the RSLR induced sediment demand of the intertidal areas are investigated applying the 2DH modelling approach of Delft3D.

7.2   Nourishment amounts

Table 7-1 shows volume and area change of the inlet elements after 50 years evolution starting from the initial bed (i.e. tidal established morphology) and two RSLR scenarios; No RSLR and RSLR of 10 mm/year (see *section 4.4.1*). Results show decreasing of the ebb-tidal delta and tidal flat volumes while increasing the channel volume of the non-zero RSLR scenario. Thus, an additional sediment demand exists in the inlet, which can for instance be provided on the ebb-tidal delta by means of nourishment. Nourishment schemes are generally designed for short periods (e.g. 5 years) (Hanson et al., 2002). However, this analysis considers a decadal application in order to counteract the RSLR induced sediment demand.

Inlet parameter	Initial bed	50-year evolution from the initial bed		Net effect of RSLR
		No RSLR	RSLR (10 mm/yr)	
Ebb-tidal delta volume (Mm^3)	169	175	145	-30
Basin sand volume (Mm^3)	33	93	137	+44
Channel: Area (km^2)	224	184	246	+62
Volume (Mm^3)	454	418	592	+174
Tidal flat: Area (km^2)	59	93	29	-64
Volume (Mm^3)	24	47	5	-42

Table 7-1 Predicted evolution of inlet elements after 50 years from the initial bed (see Figure 4-6)

The nourishment amount was based on two criteria (i.e. tidal flats and entire basin evolution). In the first case, the nourishment amount is estimated by the sediment demand of the tidal flats due to the RSLR and losses of the nourishment scheme (*Eq.* 7-1).

Nourishment amount = RSLR induced demand + losses   7-1

The RSLR induced sediment demand on the tidal flats is (47-5) Mm^3 during the 50 years evolution (Table 7-1). Nourishment schemes assume 10 – 20 % of increase on the required sediment volume due to losses (Hanson et al., 2002). Thus, the necessary nourishment amount considering 20% of loss is (47-5)×1.2/50 Mm^3/year ~ 1.0 Mm^3/year.

The second criterion uses the sediment requirement to elevate the entire basin bed by the rate of RSLR. Then, the annual nourishment amount is (224+59) km^2×10 mm/year ~ 3 Mm^3/year.

7.3   Nourishment application

7.3.1   General

The estimated nourishment amounts (i.e. 1.0 and 3.0 Mm^3/year) were applied under three strategies on the initial bed. In the *Strategy 1*, a polygon is defined enclosing the area of interest and the nourishment amount is uniformly applied in this area. This strategy was employed on the ebb-tidal delta, inlet gorge and basin areas. In the *Strategy 2*, the polygon encloses the entire area of the inlet element (i.e. ebb-tidal delta and basin). Then, the nourishment was applied only on the deep areas under two alternatives, 1) increased rates of nourishment and 2) finer sediment fractions compared to the local bed. The *Strategy 3* applied nourishment on the channel edges.

Simulations were undertaken for 50 years with the RSLR of 10 mm/year and tidal boundary forcing only based on the tidal established morphology (see *Chapter 4*).

7.3.2   Strategy 1: Uniform nourishment

Nourishing the ebb-tidal delta

Figure 7-1 shows the sub-divisions of the ebb-tidal delta area; a) Channels (C1 and C2) and Shoals (S1, S2, S3 and S4), b) Rectangular areas (R1, R2, R3, R4 and R5, each area is about 30% of the ebb-tidal delta area), and c) ebb-tidal delta area. These areas were separately simulated applying the nourishment rate of 1.0 Mm^3/year to analyse the most effective area in terms of the tidal flat evolution.

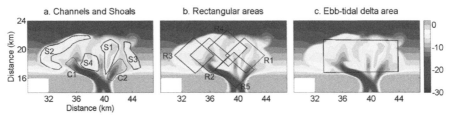

Figure 7-1 Selected nourishment areas on the ebb-tidal delta; Channel and Shoals (a), Rectangular areas (b) and Ebb-tidal delta area (c).

a. Channels and Shoals

Sediment transport through the inlet gorge showed that all nourishment areas except C1 result in almost similar landward transport as in case of the RSLR only. C1 indicated marginally higher transport ($\sim$ 2 Mm^3) being located on the main channel. Figure 7-2 shows corresponding volume changes of the ebb-tidal delta and basin. All nourishment simulations initially show similar volume change due to the quasi-flood dominant condition in the inlet (see *section 4.5.2*). The RSLR only scenario results in the strongest erosion on the ebb-tidal delta. Applying nourishment on the shoal areas indicates comparatively lower erosion compared to that of the channels. The western channel (C1) shows strong erosion on the ebb-tidal delta as evidence of the sediment transport due to the strong flood flow in this channel. The basin volume suggests that the sediment import marginally depends on the nourishment areas and there is no considerable increase in comparison to the RSLR only.

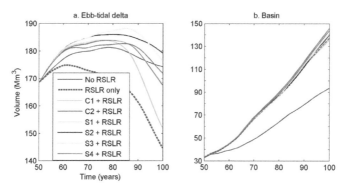

Figure 7-2 50-year evolution applying nourishment on Channels and Shoals; Ebb-tidal delta volume (a) and Basin volume (b)

b. Rectangular areas

Applying the rectangular areas, the model was simulated with increased nourishment rates (i.e. 1, 2, 3 Mm^3/year). The area R5 resulted in the highest erosion on the ebb-tidal delta and the highest sedimentation in the basin. However, this analysis also hardly showed tidal flat growth instead of sediment accumulation on the ebb-tidal delta.

c. Ebb-tidal delta area

The estimated nourishment rate of 3 Mm^3/year was applied on the representative ebb-tidal delta area. Results showed strong increase of the ebb-tidal delta volume compared to the previous two cases. Furthermore, the higher landward transport increased the basin volume (~10 Mm^3) whereas the tidal flat evolution was nearly similar to the RSLR only.

These results indicate that the predicted tidal flat evolution shows hardly any effect of the nourishment and the increased nourishment rates resulted in strong sediment accumulation on the ebb-tidal delta. Thus, the ebb-tidal delta nourishment compensates the RSLR induced sediment erosion on the ebb-tidal delta itself while hardly contributing sediment to the tidal flats.

Nourishing the inlet gorge

Two nourishment rates (i.e. 3 and 5 Mm^3/year) were undertaken on the inlet gorge. Figure 7-3 shows the volume evolution of the inlet areas during the 50-year period. A strong volume increase of the ebb-tidal delta (~100 Mm^3) is found under 5 Mm^3/year compared to the RSLR only. The inlet volume also shows the similar volume change of these two simulations on the final predicted beds. There is no such evolution in case of the basin volume implying lower sediment import into the basin. However, the basin volume increase is significant (~ 40 Mm^3) compared to nourishing the ebb-tidal delta.

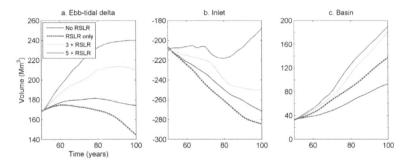

Figure 7-3 Volume change of 50-year evolution applying nourishment on the inlet gorge; Ebb-tidal delta (a), Inlet (b) and Basin (c)

The basin evolution was further analysed in terms of channel and tidal flat volumes. The No RSLR scenario resulted in almost constant channel volume while the RSLR only shows the highest channel volume. Predicted channel volumes with the nourishment are found in between these two cases such that the lower the nourishment rate the higher the channel volume. This is evidence of increased sediment import into basin as the rate of nourishment increases. However, the tidal flat evolution shows marginal growth implying that a large amount of the imported sediment remains in the channels rather than moving to the tidal flats.

Nourishing the back barrier basin

Figure 7-4 shows the selected areas in the basin nourishment (3 Mm^3/year), 1). highly dynamic area adjacent to the inlet (black-polygon, ~ 1/10th of the basin area), 2). along the main basin channel (red-polygon).

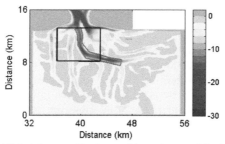

Figure 7-4 Selected areas of basin nourishment; around the inlet (black-polygon) and along the main channel (red-polygon)

Under these two cases, the ebb-tidal delta and inlet volumes show different evolutions implying different sediment distributions in the inlet (Figure 7-5). In contrast, the basin evolution is more or less similar and indicates a strong increase (~ 75 Mm^3) in comparison to the previous analyses. Furthermore, the basin channel volume is also similar in both cases and there is still marginal tidal flat growth on the final predicted beds.

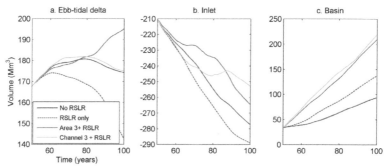

Figure 7-5 Volume change of 50-year evolution applying nourishment on the basin; Ebb-tidal delta (a), Inlet (b) and Basin (c)

7.3.3   Strategy 2: Deep-area nourishment

Deep area nourishment with different rates: 3, 5, 7 and 9 Mm^3/year

The *Strategy 2* applied nourishment in deep areas only (i.e. referring to a threshold depth) within the selected area. Initially, the ebb-tidal delta was undertaken with a threshold depth of 6 m (i.e. averaged depth of the ebb-tidal delta area). Figure 7-6 shows net erosion/accretion patterns of the predicted morphologies (i.e. [nourishment+RSLR] scenario – RSLR only scenario). Results indicate that the higher the nourishment rate the higher the accretion on the ebb-tidal delta. The main inlet channels gradually disappear as the rate of nourishment increases. This is probably due to the strong decrease of the channel flow as a result of dumping sediment on the deep areas only. This phenomenon was further evident with the basin channel and tidal flat evolutions. The channel volume decreases with the rate of nourishment while the tidal flats exhibit an evolution similar to the case with RSLR only. Thus, weak sediment supply from channels to tidal flats is apparent as found under the *Strategy 1*.

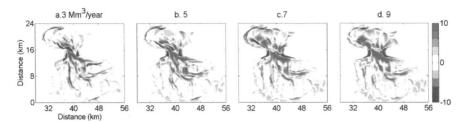

Figure 7-6 Net erosion/accretion pattern ([nourishment+RSLR]-RSLR only) after 50-year evolution of deep-area nourishment; 3 Mm³/year (a), 5 (b), 7 (c) and 9 (d)

Applying the deep-area nourishment on the inlet gorge resulted in strong accretion closing the inlet gorge. Then, the basin area was undertaken using a threshold depth of 0.5 m referring to the water surface (i.e. sediment dumps on the entire basin during HW and only in the channels during LW).

Figure 7-7 shows the volume evolution of the ebb-tidal delta, inlet and basin. The highest nourishment rate (9 Mm³/year) increases the ebb-tidal delta and basin volume by a factor of ~2 compared to the RSLR only. In this case, the ebb-tidal delta has a very shallow channel pattern while the basin channels appear to have disappeared. The inlet shows strong decrease in erosion based on the rate of nourishment (Figure 7-7*b*). However, the nourishment rate of 7 Mm³/year shows higher erosion compared to that of 5 Mm³/year. Furthermore, the highest nourishment rate tends to develop similar erosion as in case of 5 Mm³/year. These are probably due to the main inlet channel configuration on the predicted beds. A westward oriented deep channel (in 7 Mm³/year case) results in stronger erosion of the inlet than a shallow two-channel system (in 5 and 9 Mm³/year cases).

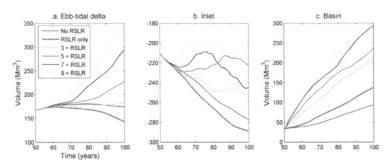

Figure 7-7 Volume evolutions of different deep-area nourishment rates on the basin; Ebb-tidal delta (a), Inlet (b) and Basin (c)

In all cases, the tidal flat evolution shows marginal growth compared to the RSLR only. This also implies strong decrease of the channel flow as a result of strong sediment filling due to the increased nourishment rates.

Deep area nourishment with fine-sediment fractions (D_{50}): 0.20, 0.10 and 0.05 mm

The second alternative used fine-sediment fractions with the rate of 3 Mm³/year. Predicted bed evolutions imply that the lower the D_{50} the higher the shoal areas in the basin and the stronger the channel pattern (Figure 7-8). This is probably due to increasing the sediment transport capacity by decreasing the D_{50}. Further, the ebb-tidal delta and basin volumes increase while decreasing the inlet volume. The tidal flat volume increases and the channel volume decreases indicating increased sediment supply into the basin. However, it can not still enable the RSLR induced sediment demand on the tidal flats.

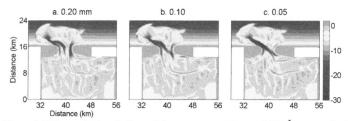

Figure 7-8 50-year bed evolutions of deep-area nourishment 3 Mm³/year on the basin with fine-sediment fractions (D_{50}); 0.20 mm (a), 0.10 mm (b) and 0.05 mm (c)

The previous analyses qualitatively imply,

1. RSLR (10 mm/year) results in positive accommodation space in the basin.
2. Increased rate of nourishment results in strong sediment filling in the channels rather than increasing sediment supply to the tidal flats.
3. Nourishment of ebb-tidal delta, inlet gorge or channels can not supply the RSLR induced sediment demand on the tidal flats.
4. Nourishment decreases the RSLR induced erosion on the ebb-tidal delta.

Next, the rate of RSLR was gradually decreased down to 5 mm/year (10, 9, 8 mm/year...etc) while applying the nourishment rate of 3 Mm³/year (D_{50} = 0.20 mm). Resulting evolutions are discussed only for the highest and the lowest RSLR scenarios.

Figure 7-9 shows the evolution of channel and tidal flat volumes. Applying nourishment and RSLR (5 mm/year, red-dash line) tends to result in similar channel volume as in case of the No RSLR (solid-black line). Similar characteristics are found between the nourishment and RSLR (10 mm/year, green-dash line) and the RSLR only (5 mm/year, red-solid line). Furthermore, the channel volume appears to be proportional to the rate of RSLR. These are clearly evidence of the channel volume sensitivity to both nourishment and RSLR rates. The tidal flat volume shows marginal effect of the nourishment (Figure 7-9*b*). Thus, it is very sensitive only to the RSLR of which 5 mm/year results in comparatively stable tidal flat evolution (see *section 4.5*). These results also imply strong sediment accumulation in the channels rather than tidal flat growth.

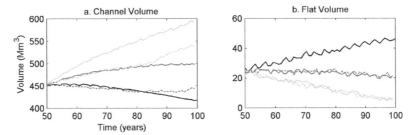

Figure 7-9 Evolution of channel (a) and tidal flat (b) volume of different RSLR scenarios with nourishment rate of 3 Mm³/year; No RSLR (black-line), 5 mm/year (red-line), 10 mm/year (green-line) and dash-lines indicate nourishment with corresponding RSLR

So far, the model predictions showed weak sediment supply to the tidal flats based on a *moving reference* (i.e. varying MSL with the rate of RSLR). Thus, the bed evolution is further investigated with a *constant reference* (i.e. initial MSL as in No RSLR) to find the actual sediment supply to the tidal flats. Following parameters are herein employed,

Volume change of the inlet elements (ΔV),

$$\Delta V_i = V_{Final,i} - V_{Initial,i} \qquad\qquad\qquad 7\text{-}2$$

where, i = Ebb-tidal delta, Inlet, Basin, Basin channel, Basin tidal flat; $V_{Initial,i}$, inital element volume; $V_{Final,i}$, predicted element volume on the final bed.

Difference in average tidal flat height (Δh),

$$\Delta h = h_{Final} - h_{Initial} \qquad\qquad\qquad 7\text{-}3$$

where, $h_{Initial}$ and h_{Final} are initial and final predicted tidal flat heights respectively.

Model ID	RSLR (mm/yr)	Nourishment rate (Mm³/yr)	Reference level for tidal flats/channels	Ebb	Inlet	Basin	Overall	Basin channel	Basin tidal flat	$h_{Initial}$	h_{Final}	Δh
				\multicolumn{6}{}{Volume change (Mm³), ΔV_i}				\multicolumn{3}{}{Flat height (m)}
M1	No	No	Constant	6	-67	60	-1	-39	18	0.44	0.48	0.04
M2	No	3	Constant	42	-19	128	151	-104	20	0.44	0.48	0.04
M3	5	No	Moving	-2	-85	88	1	44	-3	0.42	0.34	-0.08
M4	5	3	Moving	25	-21	147	151	-11	-3	0.42	0.32	-0.10
M5	**5**	**No**	**Constant**	**-2**	**-85**	**88**	**1**	**-54**	**31**	**0.44**	**0.54**	**0.10**
M6	**5**	**3**	**Constant**	**25**	**-21**	**147**	**151**	**-107**	**36**	**0.44**	**0.55**	**0.11**
M7	10	No	Moving	-25	-78	104	1	137	-20	0.42	0.17	-0.25
M8	10	3	Moving	14	-35	171	150	84	-18	0.42	0.21	-0.21
M9	**10**	**No**	**Constant**	**-25**	**-78**	**104**	**1**	**-65**	**45**	**0.44**	**0.63**	**0.19**
M10	**10**	**3**	**Constant**	**14**	**-35**	**171**	**150**	**-119**	**48**	**0.44**	**0.63**	**0.19**

Table 7-2 Predicted element evolution after 50 years applying different nourishment/RSLR scenarios with respect to moving/constant references

Table 7-2 shows the estimated element evolution based on the constant and moving references. In case of the constant reference, the basin volume change (ΔV_{Basin}) tends to agree with the volume changes of basin channels and tidal flats ($\Delta V_{Basin,cf}$).

$$\Delta V_{Ba\sin,cf} = \Delta V_{Ba\sin\,tidal\,flats} - \Delta V_{Ba\sin\,channel}$$
7-4

The constant reference results in decreasing the channel volume while importing sediment into the basin. Thus, the decreased channel volume indicates the amount of sediment import into the basin. However, this is not the case of the moving reference because the channel volume is mainly governed by the moving LW level. The $h_{Initial}$ of the constant reference is slightly higher (~0.02 m) than that of the moving reference. This is due to the selection of time windows for the bed level and water level results of the simulations (i.e. averaged LW and HW is estimated for the time window of bed level results to analyse the element evolution).

The nourishment application only (M2) shows decreasing the channel volume ($\Delta V_{Basin\ channel}$) whereas the tidal flat evolution (Δh) is similar to the No nourishment and No RSLR case (M1). The channel volume is very sensitive to both nourishment and RSLR rates (see M3 and M4; M7 and M8). The model results of the RSLR only indicate an accelerated sediment import into the basin (see Δh in M5 and M9, *constant reference*). However, it is not sufficient to develop the tidal flats as the MSL also increases (see Δh in M3 and M7, *moving reference*). The M6 and M10 (*constant reference* with nourishment and RSLR) imply that if the estimation is based on the initial MSL, the channel volume significantly decreases while growing the tidal flats as in case of M5 and M9. Thus, the nourished sediment has a marginal contribution on the tidal flat growth (see $\Delta V_{Basin\ tidal\ flats}$ in M5 and M6; M9 and M10). Therefore, the results suggest, the nourishment schemes can not enable the tidal flat growth with the rate of RSLR (see the cases of *moving reference*).

7.3.4   Strategy 3: Channel-edge nourishment

The *Strategy 3* investigates channel nourishment (i.e. depth from 3 to 10 m) to supply the RSLR induced sediment demand on tidal flats in a 5-year period. Channel nourishment is likely to increase the intertidal areas (Camille, 2010; De Ronde et al., 2009). Using a 2DH modelling approach, Das (2010) discussed the sediment interaction between channels and tidal flats based on *Galgeplaat* in the Eastern Scheldt, the Netherlands. Results indicate sediment supply from channels to tidal flats under strong tidal currents and vice versa during storm conditions.

Initially, this strategy was applied on the ebb-tidal delta channels (i.e. western and eastern, see Figure 7-1*a*). Resulting evolution showed weak tidal flat growth. Then, the inlet gorge channels were nourished. Only the western channel resulted in comparatively increase of the tidal flat evolution (Model C1 in Table 7-3). Finally, the main basin channel was investigated using three alternatives, 1). both west and east channel edges applying 3.0 Mm^3/year (C20), 2). both west and east channel edges applying 1.5 Mm^3/year (C21) and 3). only western channel edge applying 3.0 Mm^3/year (C22). The M9 and M10 (from Table 7-2) show the results of the first 5-year evolution based on the constant reference. The ΔV_{diff} refers to the difference of ΔV_{Basin}- $\Delta V_{Basin,cf}$ (see *Eq.* 7-4).

Model ID	RSLR (mm/yr)	Nourishment rate (Mm^3/yr)	Volume change (Mm^3), ΔV_i							Flat height (m)		
			Ebb	Inlet	Basin	Net change	Basin channel	Basin tidal flat	ΔV_{dif}	$h_{Initial}$	h_{Final}	Δh
M9	10	No	4.2	-9.6	5.3	-0.2	-1.1	2.7	-1.4	0.44	0.43	-0.01
M10	10	3.0	4.5	-9.7	20.2	15.0	-15.3	2.8	-2.1	0.44	0.44	0
C1	10	3.0	7.1	2.1	5.0	14.2	-2.1	2.6	0.3	0.44	0.43	-0.01
C20	10	3.0	5.0	-7.9	30.7	27.7	-26.1	4.2	0.3	0.44	0.44	0
C21	10	1.5	4.6	-8.9	17.6	13.2	-14.0	3.2	0.4	0.44	0.43	-0.01
C22	10	3.0	4.6	-8.8	17.5	13.3	-13.7	3.5	0.3	0.44	0.44	0

Table 7-3 Predicted element evolution after 5 years applying nourishment on channel edges based on the constant reference

The ΔV_{diff} has been significantly decreased after applying a small time-window for the bed level results (see M9, M10 with others in Table 7-3). The C1 and C21 show higher reduction of the tidal flat height (Δh) while stable evolution is found in the C20 and C22. The C20 has the highest growth of the tidal flats whereas the applied nourishment amount is about twice (i.e. both channel edges) than that of the C22. Thus, the tidal flat growth appears to be comparatively stable under the 5 years period.

These results indicate that the mechanism of sand sharing between the different elements (i.e. ebb-tidal delta, channel and tidal flats) as assumed in early studies using e.g. ASMITA, is not reproduced by Delft3D. Before discussing this further, it is worth to illustrate the difference by carrying out some similar simulations in the ASMITA model.

7.4   Nourishment with ASMITA

Three-element version of the ASMITA model is applied (i.e. ebb-tidal delta, channels and tidal flats, see *section 6.2.2*). Channel volume initially decreases due to nourishment in the basin channels. This phenomenon is used to implement nourishment in the ASMITA model by decreasing the initial channel volume similar to the nourishment amount. Present analysis undertakes the case of 3 Mm^3/year nourishment in a 50 years period (i.e. 3×50 Mm^3). Applying nourishment annually does not provide more insight of the element evolutions. Therefore, a hypothetical scheme is used to investigate the tidal flat evolution, i.e. 150 Mm^3 initial reduction of channel volume. Initially, evolution of the tidal established morphology as in *section 7.3* was investigated from 50 to 100 years under this hypothesis. Then, the Ameland inlet (i.e. 2004 data) was modelled from 2004 to 2054.

7.4.1   Schematised inlet

Figure 7-10 shows the volume evolution of channels and tidal flats. The initial channel volume of the nourishment scenarios (red-dash: 5 mm/year and green-dash: 10 mm/year) is 150 Mm^3 lower than the other cases. These two models predict strong increase of the channel volume to reach equilibrium state. The tidal flat evolution is also stronger with the nourishment indicating increase sediment supply from channels to tidal flats. In all cases, the higher the rate of RSLR the faster the flat volume tends to be stable. This implies lower tidal flat growth under higher rate of RSLR. Finally, the flat volume of the nourishment with 5 mm/year appears to agree with that of the No RSLR (black-line). Furthermore, the nourishment with 10 mm/year shows similar tidal flat volume with the RSLR of 5 mm/year (red-line). In case of the Delft3D simulations, these characteristics were found only with the channel volume (see Figure 7-9). Therefore, results indicate different sand sharing mechanisms between inlet elements under both modelling approaches.

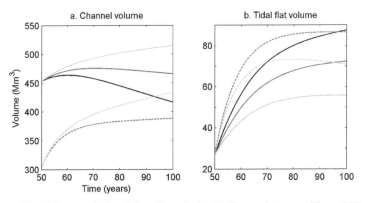

Figure 7-10 Volume evolution of the schematised inlet after applying nourishment 150 Mm^3 at the beginning, Channel (a) and Tidal flat (b); No RSLR (black-line), 5 mm/year (red-line), 10 mm/year (green-line) and dash-lines indicate nourishment with corresponding RSLR

7.4.2   Ameland inlet

Initial element parameters of the Ameland inlet are based on the 2004 measured data and referred to Table 7-4.

Description	Element	Area (10^6 m^2)	Volume (10^6 m^3)	Equilibrium coefficient
Forecasting from 2004 to 2054	Tidal flats	172	120	0.1838
	Channels	102	302	11.0
	Ebb-tidal delta	75	131	2921.6

Table 7-4 Initial conditions and equilibrium coefficients of the ASMITA run for forecasting the Ameland inlet from 2004 to 2054

Figure 7-11 shows the evolution of channel and tidal flats. In contrast to the schematised inlet, the Ameland inlet appears to be stable under No RSLR during the 50 years period (from 2004 to 2054). However, channel volume increases while decreasing the tidal flat volume under RSLR scenarios (i.e. red-solid-line 5 mm/year and green-solid-line 10 mm/year). The growth in channel volume is about 25 and 50 Mm^3 in the case of RSLR 5 and 10 mm/year respectively and the corresponding reductions in tidal flat volumes are about 40 and 80 Mm^3. Therefore, both evolutions appear to be proportional to the rate of RSLR. Channel volume under nourishment scenarios (dash-lines) shows gradual increase as in case of the RSLR only cases. On the final predicted beds, the difference of channel volume between [nourishment+RSLR] and RSLR only cases is about twice (150 Mm^3) compared to that of schematised inlet (80 Mm^3). Predicted tidal flat volumes in the nourishment cases are almost similar to that of the RSLR only cases (i.e. red-dash-line and red-line, green-dash-line and green-line in Figure 7-11*b*). This is probably due to, 1) being a stable tidal inlet and 2) the larger flat volume (~ 120 Mm^3) and lower channel volume (~ 300 Mm^3), than the schematised inlet and therefore the applied nourishment (i.e. reduced channel volume) is not sufficient to strongly increase the tidal flat growth. This was further evident by increasing the initial reduction of channel volume which resulted in slight increase in the tidal flat evolution (*not shown*).

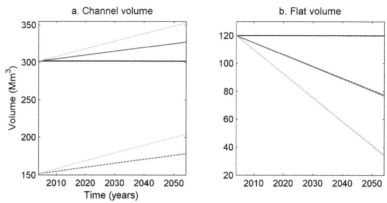

Figure 7-11 Volume evolution of the Ameland inlet (2004) after applying nourishment 150 Mm³ at the beginning, Channel (a) and Tidal flat (b); No RSLR (black-line), 5 mm/year (red-line), 10 mm/year (green-line) and dash-lines indicate nourishment with corresponding RSLR

7.5   Discussion

Under the schematised inlet, the process-based model (Delft3D) showed that the sand nourishment hardly fulfills the RSLR induced sediment demand of tidal flats. However, the predicted evolutions of the ASMITA model showed strong growth of the tidal flats in contrast to the Delft3D model. These indicate that there are different sediment distribution mechanisms between inlet elements under both modelling approaches. Applying the Ameland inlet, the ASMITA model predicted decrease of tidal flats in the nourishment scenarios. The Ameland inlet has higher volume of tidal flats and lower volume of channels in comparison to the schematised inlet. Thus, the initial characteristics of the inlet elements determine the distribution of the applied nourishment.

Results generally helped to understand the possible evolution of the inlet elements under applying nourishment to counteract the RSLR induced sediment demand. Only basin channel nourishment appears to have developed comparatively stable tidal flats in the Delft3D model. Thus, further research on this application is recommended because applying nourishment to counteract the RSLR effects is not yet comprehensively investigated.

7.6   Conclusions

Numerical experiments (Delft3D) were undertaken to investigate whether the ebb-tidal delta nourishment fulfils the RSLR induced sediment deficit of tidal inlets (i.e. erosion of ebb-tidal delta and demand on tidal flats). Initially, a schematised tidal inlet which is analogous to the Ameland inlet, was simulated for 50 years with the tidal boundary forcing only under three nourishment strategies, 1). *uniform nourishment*, 2). *deep-area nourishment* and 3). *channel-edge nourishment*. Finally, the optimal strategy was further investigated applying the ASMITA model on the schematised inlet and the Ameland inlet.

Nourishment amount was estimated based on the RSLR induced sediment demand on the tidal flats (1.0 Mm^3/year) and the entire basin sediment requirement to follow the RSLR (3.0 Mm^3/year). Different areas on the ebb-tidal delta were nourished and analysed the tidal flat evolution. *Strategy 1* hardly showed any effect of the nourishment whereas it compensates the erosion on the ebb-tidal delta itself. Increased nourishment rates resulted in strong sedimentation and disappearing the channel system. Under *Strategy 2*, applying nourishment on the ebb-tidal delta and inlet gorge indicated marginal growth of the tidal flats. This was further evident applying nourishment on the basin area with increased nourishment rates. Applying nourishment with the fine-sediment fractions resulted in comparatively increased evolution though it was inadequate to fulfill the RSLR induced sediment demand on the tidal flats. Under this strategy, strong sediment filling in the channels was observed due to increased nourishment rates which resulted in decreasing the channel flow. This further implies strong decrease of the sediment supply from channels to tidal flats. *Strategy 3* resulted in comparatively stable tidal flats by nourishing on the basin channel edges in short morphological periods (i.e. 5 years). Finally, this strategy was investigated with the ASMITA model by reducing the initial channel volume to implement nourishment application. Results showed strong growth of tidal flats under schematised inlet and strong decrease of tidal flats under the Ameland inlet. These indicate different mechanisms of sediment distribution in both models.

Results of this analysis suggest possible evolution of the inlet elements under the effect of RSLR and nourishment scenarios. At present, applying nourishment on tidal inlets to mitigate the RSLR induced sediment demand is not completely understood and therefore it is recommended to carry out further investigation on the basin channel nourishment of which the predicted tidal flat evolution appears to be stable.

Chapter 8

Conclusions and recommendations

The overarching aim of this thesis is to investigate the evolution of a tidal inlet in response to sea level rise and to evaluate the possible mitigation measures for the potential negative impacts. This objective has been achieved according to the results presented in the previous chapters.

In order to achieve the main objective, four specific research questions were formulated. Conclusions are reached by elaborating these questions. Recommendations are given for aspects needing further investigations.

8.1   Conclusions

Objective 1:   Can a process-based approach be reliably applied on the decadal time scale, and does such a model predict geometric properties similar to observations?

Present study used a schematised tidal inlet representing the Ameland inlet. Simulations were undertaken applying the process-based model Delft3D with the Realistic Analogue approach (see Roelvink and Reniers, 2011) to investigate the decadal evolution (50 years). Sensitivity of the bed evolutions was analysed in terms of different model/physical parameters. All predicted beds qualitatively agree with the major morphological features of the Ameland inlet (i.e. eastward oriented basin channel pattern, westward skewed main inlet channel and ebb-tidal delta). Effect of different *MORFAC* values was marginal on the overall bed configuration whereas the optimal match to the Ameland inlet was determined via the statistical comparison (*BSS* > 0.3). The Van Rijn (1993) transport formula resulted in strong bed evolution in comparison to the Engelund-Hansen (1967) formula. Predicted evolutions qualitatively agree with the previous research of Soulsby (1997) implying that the Van Rijn formula results in approximately twice the sediment transport rate as in the Engelund-Hansen formula. The extended morphological period up to 100 years of the latter formula indicated the same major morphological features of the Ameland inlet as in the Van Rijn formula. Resulting morphologies were very sensitive to the direction and asymmetry of the tidal boundary forcing and to the physical parameters (i.e. depth, inlet width and basin location relative to the inlet gorge). Accordingly, the present-day channel/shoal pattern of the Ameland inlet is mainly due to the west-east propagating alongshore tide with the eastward skewed basin location. Predicted inlet morphology after including the wave effect also consists of a stable two-channel system, which better resembles the Ameland inlet (*BSS* > 0.4). These results suggest that the process-based model Delft3D can potentially be applied for simulating decadal inlet evolution and predicts a geometry representative of the observations.

Objective 2:   Does a process-based approach support the assumptions applied in semi-empirical models, such as the level of tidal flats following sea level rise with some time lag, and does this still hold under accelerated sea level rise?

Responding inlet evolution to sea level rise was investigated based on an established morphology (i.e. predicted analogous bed to the Ameland inlet). Results of the process-based model (Delft3D) indicated that the existing flood dominance of the system accelerates as the rate of RSLR increases. Thus, the ebb-tidal delta erodes while the basin accretes. The erosion/accretion rates are positively correlated with the rate of RSLR. The tidal flats continued to develop under the No RSLR condition. Under the high RSLR, the tidal flats eventually drowned, implying that the system may degenerate into a tidal lagoon. The low RSLR (~5 mm/year and IPCC L+LS scenario (average ~4 mm/year)) resulted in nearly stable tidal flat evolution implying that this may be the critical RSLR condition for the maintenance of the system. In contrast, the semi-empirical model (ASMITA) showed much higher critical RSLR rate of the Ameland inlet (~10.5 mm/year). This difference indicates the uncertainties in both approaches. Under the No RSLR condition, resulting bed evolutions of these two models strongly converge after the initial adaptation periods. The semi-empirical model predicts an increase of the adaptation period as the rate of RSLR increases and thus this model demands a long morphological period (~ 100 years) to predict comparable evolution to that of the process-based model. This analysis suggested that the inlet evolutions of the process-based model strongly agree with the semi-empirical assumptions under the No RSLR condition. However, under the RSLR conditions (i.e. accelerated sea level rise), the process-based results tend to weakly agree with these assumptions probably due to strong increase of the initial adaptation period of the semi-empirical results.

Objective 3:   What are the alongshore length scales of inlet influence on adjacent coastlines, and what are the dominant processes?

Inlet effects on the adjacent coastlines and related physical processes were investigated based on the evolutions of inlet-bathymetry and no inlet-bathymetry (i.e. undisturbed coastline) imposing both tidal and wave boundary forcings. Resulting sediment transport pattern indicated a tidal-flow bypassing system and the bed evolution showed onset of a cyclic behaviour of the main inlet channel. Both these processes are typical to the Ameland inlet. Sediment circulation of the inlet showed that the ebb-tidal delta has a significant influence on the eastern coastline as the west-east directed alongshore transport is strongly interrupted. Inlet effect on the barrier islands tends to develop 's' signature of the coastline behaviour as typically found at tidal inlets (Galgano, 2009a,b). The non-zero RSLR scenario showed strong bed evolution and in turn inlet effect around the inlet due to the increased MSL which results in wave breaking close to the coastline on the ebb-tidal delta in comparison to the No RSLR. Further, the RSLR effect on the bed evolution was marginal on the east compared to that on the west. This is mainly due to the strong wave generated currents on the west and the wave sheltering effect of the ebb-tidal delta on the east (i.e. northwesterly dominant wave direction). Predicted coastline effects appear to be overestimated (i.e. ~15 km on the west and 25 km on the east). This is probably due to the model schematisations and the analysis of a single inlet of a series of inlets where the sediment exchange of the adjacent basins was not implemented. Average length of the barrier islands of the Ameland inlet is about 25 km. Therefore, the inlet effect is likely to exist on the entire barrier island coastlines due to the adjacent inlets. This analysis identified the following physical processes relating to the inlet effect on adjacent coastlines; *sediment bypassing system, cyclic behaviour of the main inlet channel, wave breaking on the ebb-tidal delta* and *wave sheltering effect*.

Objective 4: Can nourishment of the ebb-tidal delta be an effective means of feeding a tidal inlet and if so, where on the ebb-tidal delta is the most effective nourishment location?

Sand nourishment on ebb-tidal delta was selected as the mitigation measure to counteract the negative impacts of RSLR and undertaken based on three strategies (i.e. *uniform nourishment, deep-area nourishment* and *channel-edge nourishment*) applying on the established morphology. The optimal strategy of the Delft3D model was further investigated employing the ASMITA model also. The nourishment amount was estimated based on the model predicted evolutions. Different areas on the ebb-tidal delta were nourished and analysed for the bed evolution. Resulting tidal flat evolution hardly showed any effect of the nourishment whereas strong sediment accumulation was evident on the ebb-tidal delta after increasing the rate of nourishment in all three strategies. Results suggested that the ebb-tidal delta nourishment can not enable the RSLR induced sediment demand of the tidal flats and it only compensates the sediment demand on the ebb-tidal delta itself. Thus, sand nourishment in the basin area was investigated. Application of the finer sediment fractions resulted in comparatively increased transport to the tidal flats implying their marginal growth whereas the basin channel system disappeared with the higher rate of nourishment. Nourishing on the basin channel edges indicated comparatively more stable tidal flat evolution than other two strategies. Finally, this strategy was investigated with the ASMITA model in terms of reducing the initial channel volume to implement the effect of nourishment. Results showed strong evolution of the tidal flats in contrast to the Delft3D results. This is expected as the initial channel volume is lower than the equilibrium channel volume. Therefore, these results indicate different mechanisms of sediment distribution in both models. Present analysis shows possible evolution of the inlet elements under the effect of RSLR and nourishment implying that the sand nourishment hardly fulfills the RSLR induced sediment deficit of tidal flats.

8.2   Recommendations

Selection of MORFAC to analyse the morphological evolution

The morphological evolution of the process-based model Delft3D is explicitly based on the *MORFAC* which depends on the hydrodynamic/physical characteristics of the area of interest. The lower the *MORFAC* value the better the evolution in terms of the numerical stability (Lesser et al., 2004). So far, there is no *a priori* determination to select the highest *MORFAC* for a given study area. Thus, the present study used two different *MORFAC* approaches (i.e. one single *MORFAC* (200) and time varying *MORFACs* (10, 20, 50 and 200, see *Chapter 4*) where smaller *MORFACs* gradually change to larger *MORFACs* during the morphological period) to find the optimal *MORFAC* that gradually decreases the strong morphological shock of the schematised flat bed and develops an analogous bed to the Ameland inlet. Both approaches resulted in similar bed evolutions with respect to the major morphological features of the Ameland inlet. However, the statistical comparison suggested that the single *MORFAC* develops the closest match to the Ameland inlet. This is probably due to, 1) initial flat bed resulting to rapid morphological changes, 2) uncertainties related to the *MORFAC* approach or 3) inherent unpredictability of the smaller patterns. Therefore, it is strongly recommended to undertake a thorough investigation on the *MORFAC* selection in order to find the most desirable *MORFAC* for the area of interest.

Prediction of critical RSLR of the semi-empirical model (ASMITA) and process-based model (Delft3D)

Semi-empirical model suggested critical RSLR of ~10.5 mm/year of the Ameland inlet whereas it is ~5 mm/year of the process-based model. These critical RSLR rates are based on the tidal flat evolution which was analysed using the same criterion in both approaches. The semi-empirical model predicted the equilibrium inlet evolution in terms of the element volumes while assuming constant element areas. These results showed strong initial adaptation period that tends to increase as the rate of the RSLR increases. In contrast, the process-based model predicted comparatively gradual evolution of element volumes and areas based on the inlet hydrodynamic and the sediment transport characteristics. Therefore, the tidal flat evolution is described by the time varying area and volume. In both approaches, the model results strongly converge in case of the No RSLR and partially converge in case of the non-zero RSLR scenarios. Accordingly, the difference of the critical RSLR rates is embedded with the model uncertainties in both approaches. Therefore, it is recommended to do further research in order to reduce these uncertainties governing the physical mechanisms of tidal inlet response to RSLR.

Can nourishment fulfil the RSLR induced sediment deficit of tidal flats?

Present study used sediment nourishment to supply the RSLR induced sediment demand of tidal inlets. The Delft3D model predicted hardly any growth of tidal flats by nourishing on the ebb-tidal delta and inlet gorge. However, applying nourishment on the basin channel edge showed comparatively increase of tidal flats though it is inadequate to fulfill the RSLR induced sediment demand. This strategy was further investigated with the ASMITA model. Predicted evolution indicated strong growth of the tidal flats in contrast to the Delft3D model. Therefore, both models predicted more fruitful results in terms of the tidal flat evolution under nourishment of the basin channel edge. Further research on this application is thus recommended as the nourishment to counteract the RSLR effects is not yet comprehensively investigated.

8.3　Concluding remark

This research investigated responding evolution of a tidal inlet to RSLR applying a process-based model (Delft3D) on a schematised tidal inlet which optimised a number of sensitivity experiments rather than employing a measured inlet bathymetry. Predicted 50-year bed evolution resulted in higher resemblance with the Ameland inlet bed (i.e. *BSS* > 0.3, tidal boundary only; *BSS* > 0.4, tidal and wave boundary forcings) increasing the reliability of this approach. Thus, applying the future RSLR scenarios on the predicted morphology apparently provides representative evolution to the Ameland inlet. The primary conclusions of this research are, 1). the decadal evolution of the process-based model resulted in a representative geometry as observed in the data, 2). the critical RSLR rate of the inlet is about 5 mm/year (Delft3D) and 10 mm/year (ASMITA), 3). the results of the Delft3D model are likely to agree with the equilibrium assumptions of the semi-empirical model (ASMITA), 4). the inlet effect is stronger on the eastern coastline, and 5). the ebb-tidal delta nourishment is unlikely to be an effective mitigation measure to supply the RSLR induced sediment deficit of tidal inlets.

More research focusing on the quantification of the physical and socio-economic impacts of RSLR on these systems is needed to develop effective and timely adaptation strategies enabling at least the partial preservation of bio-diversity and local communities in these regions.

List of Symbols

Symbol	Units	Meaning
$\bar{c}$	kg/m^3	depth averaged sediment concentration
$\bar{c}_{eq}$	kg/m^3	depth-averaged equilibrium concentration
c_f	-	friction coefficient
C	m$^{1/2}$/s	Chèzy coefficient
d_{50}	m	mean grain diameter
D_H	m^2/s	horizontal dispersion coefficient
D_*	-	non-dimensional particle diameter
f_{bed}	-	calibration factor
f_c'	-	current-related friction factor
f_{cor}	-	Coriolis parameter
g	m/s^2	gravitational acceleration factor
h	m	water depth
M_{tot}	m^3/year	alongshore transport
r	year	stability number
s	-	relative sediment density
S_b	kg/m/s	bed load transport
S_b'	kg/m/s	corrected transport in longitudinal direction
$S_{b,n}$	kg/m/s	additional bed load transport vector
S_x, S_y	m^2/s	sediment transport components
T_a	-	non-dimensional bed shear stress
T_d	m/s	deposition or erosion rate
T_s	s	adaptation time-scale
T_{sd}		analytical function of shear velocity
$\bar{u}, \bar{v}$	m/s	depth averaged velocity in x and y directions
u_b	m/s	near bed fluid velocity vector
$u_{b,cr}$	m/s	critical near-bed fluid velocity
u_*	m/s	bed shear velocity
w	m/s	sediment fall velocity
z_b	m	bed level
z_{Falt}	m	bed level of flat schematised bathymetry
$z_{measured}$	m	bed level of measured bathymetry
$z_{modelled}$	m	bed level of model predicted bathymetry
α	-	calibration coefficient
α_{bn}	-	transverse bed slope factor
α_{bs}	-	longitudinal bed slope factor
α_s	-	correction factor
Δ	-	relative density
Δz	m	bed level changes

Symbol	Units	Meaning
Ω	m^3	tidal prism
ε	-	bed porosity
$\o$	deg.	angle of repose
φ	deg.	phase relative to a fixed point in time and space
υ	m^2/s	horizontal eddy viscosity
η	-	relative availability of sand at bottom
η	m	water level
$\hat{\eta}$	m	amplitude of tidal wave
ρ_s	kg/m^3	density of sediment
ρ_w	kg/m^3	water density
$\tau_{b,cr}$	N/m^2	critical bed shear stress for initiation of sediment transport
ω	deg./hour	angular frequency
ζ	m	water level

Bibliography

Anonymous, 1990. A New Coastal Defence Policy for the Netherlands. Ministry of ransport and Public Works, Rijkswaterstaat, 103.

Aubrey, D.G., Speer, P.E., 1985. A Study of non-linear tidal propagation in shallow Inlet/Estuarine Systems, Part I: Observations, *Journal of Estuarine, Coastal and Shelf Science* 21, 185-205.

Bagnold, R.A., 1966. An approach to the Sediment Transport Problem from General Physics. Geol. Surv. Prof. Paper vol. 422-I, US Geological Survey, DOI, USA.

Beets, D.J., and Van der Speck, J.F., 2000. The Holocene evolution of the barrier and the back-barrier basins of Belgium and the Netherlands as a function of late Weichselian morphology, relative sea-level rise and sediment supply, *Netherland Journal of Geoscience*, 79, 3-16.

Bindoff, N.L., J. Willebrand, V. Artale, A, Cazenave, J. Gregory, S. Gulev, K. Hanawa, C. Le Quéré, S. Levitus, Y. Nojiri, C.K. Shum, L.D. Talley and A. Unnikrishnan, 2007: Observations: Oceanic Climate Change and Sea Level. In: Climate Change 2007: The Physical Science Basis. Contribution of Working Group I to the Fourth Assessment Report of the Intergovernmental Panel on Climate Change (IPCC) [Solomon, S., D. Qin, M. Manning, Z. Chen, M. Marquis, K.B. Averyt, M. Tignor and H.L. Miller (eds.)]. Cambridge University Press, Cambridge, United Kingdom and New York, NY, USA, 409-416.

Bird, E.C.F., 1985. Coastline Changes – A Global Review, Wiley Interscience, Chichester, 219.

Booij, N., Ris, R.C., Holthuijsen, L.H., 1999. A third generation wave model for coastal regions, Part I, Model description and validation, *Journal of Geophysical Research*, 104, C4, 7649-7666.

Boon, J.D. and Byrne, R.J., 1981. On basin hypsometry and the morphodynamic response of coastal inlet systems, *Marine Geology* 40, 27-48.

Bruun, P., 1986. Morphological and navigational aspects of tidal inlets on littoral drift shores, *Journal of Coastal Research*, 2, 123-143.

Bruun, P., Gerritsen, F., 1959. Natural by-passing of sand at coastal inlets, *Journal of Waterways and Harbours Division*, Proc. ASCE, 75-107.

Bruun, P., Gerritsen, F., 1960. Stability of Coastal Inlets, North Holland Publishing Co., Amsterdam, The Netherlands.

Camille, B., 2010. Monitoring of the ecological and morphological changes of the sand nourishment pilot on the Galgeplaat, Aquatic ecotechnology, Bachelor of Water Management, Hogeschool Zeeland.

Capobianco, M., Hanson, H., Larson, M., Steetzel, H., Stive, M.J.F., Chatelus, Y., Aarninkhof, S., Karambas, T., 2002. Nourishment design and evaluation: applicability of model concepts, *Coastal Engineering*, 47, 113-135.

Castelle, B., Bourget, J., Molnar, N., Strauss, D., Deschamps, S., Tomlinson, R., 2007. Dynamics of a wave-dominated tidal inlet and influence on adjacent beaches, Currumbin Creek, Gold Coast, Australia, *Coastal Engineering*, 54, 77-90.

Cheung, K.F., Gerritsen, F., Cleveringa, J., 2007. Morphodynamics and sand bypassing at Ameland Inlet, The Netherlands, *Journal of Coastal Research*, 23(1), 106-118.

Cleveringa, J. and Oost, A. P. 1999. The fractal geometry of tidal-channel systems in the Dutch Wadden Sea, Geologie en Mijnbouw, 78, 21-30.

Coe, A.L., and Church, K.D., 2003. The sedimentary record of sea-level change, Cambridge University Press, 58-61.

Cowell, P.J., Stive, M.J.F., Niedoroda, A.W., Swift, D.J.P., De Vriend, H.J., Buijsman, M.C., Nicholls, R.J., Roy, P.S., Kaminsky., G.M., Cleveringa, J., Reed, C.W., and De Boer, P.L., 2003. 'The Coastal-Tract (Part 2): Applications of Aggregated Modelling to Lower-order Coastal Change', *Journal of Coastal Research* 19 (4), 812-827.

Das, I., 2010. Morphodynamic modelling of the Galgeplaat, MSc thesis, Delft University of Technology.

Dastgheib, A., Roelvink, J.A., Wang, Z.B., 2008. Long-term process-based morphological modelling of the Marsdiep Tidal Basin, *Marine Geology*, 256, 90-100.

Dean, R.G., and Work, P.A., 1993. Interaction of Navigation Entrances with Adjacent Shorelines. *Journal of Coastal Research,* 18: 91-110.

De Fockert, A., Stive, M.J.F., Wang, Z.B., Luijendijk, A.P., De Ronde, J.G., De Boer, G.J., 2008. Impact of Relative Sealevel Rise on the Amelander Inlet Morphology, MSc thesis, Technical University of Delft.

Denys, L., and Baeteman, C., 1995. Holocene evolution of relative sea level and local mean high water spring tides in Belgium – a first assessment, *Marine Geology*, 124, 1-19.

De Jong, F.; Bakker, J.F.; van Berkel, C.J.M.; Dankers, N.M.J.A.; Dahl, K.; Gätje, C.; Marencic, H.; Potel, P., 1999. Wadden Sea Quality Status Report. Wadden Sea Ecosystem No. 9. Common Wadden Sea Secretariat, Trilateral Monitoring and Assessment Group, Quality Status Report Group. Wilhelmshaven, Germany.

De Ronde, J., Mulder, J., Ysebaert, T., Van Duren, L., 2009. Kaderplan Autonome Neerwaartse Trend, ANT Oosterschelde, Deltares report (*in Dutch*).

De Swart, H.E., Schuttelaars, H.M. and Bonekamp, J.G., 2004. Dynamics of channels and shoals on ebb-tidal deltas: the role of waves and tides, proc. of Physics of Estuaries and Coastal Seas (PECS), Mexico.

De Vriend, H.J., 1991. Mathematical modelling and large-scale coastal behaviour, Part 1: Physical processes, *Journal of Hydraulic Research*, 29, 727 – 740.

De Vriend, H.J., Capobianco, M., Chesher, T., De Swart, H.E., Latteux, B., Stive, M.J.F., 1993. Approaches to long-term modelling of coastal morphology: a review, *Coastal Engineering*, 21, 225 – 269.

Delft3D Flow User Manual
(http://delftsoftware.wldelft.nl/index.php?option=com_docman&task=cat_view&gid=39&Itemid=61)

Dissanayake, D.M.P.K., Ranasinghe. R., and Roelvink, J.A., 2009b. Effect of sea level rise in inlet evolution: a numerical modelling approach, *Journal of Coastal Research*, SI 56, Proc. of the 10th International Coastal Symposium, Lisbon, Portugal, 942- 946.

Dissanayake, D.M.P.K., Roelvink, J.A., 2007. Process-based approach on tidal inlet evolution – Part 1, Proc. 5th IAHR Symposium on River, Coastal and Estuarine Morphodynamics, The Netherlands.

Dissanayake, D.M.P.K., Roelvink, J.A., Ranasinghe, J.A., 2009a. Process-based approach on tidal inlet evolution – Part II, Proc. International Conference in Ocean Engineering, Chennai, India. CD Rom.

Dissanayake, D.M.P.K., Van der Wegen, M., and Roelvink, J. A, 2009c. Modelled channel pattern in schematized tidal inlet, *Coastal Engineering*, 56, 1069 – 1083.

Dissanayake, D.M.P.K., Roelvink, J.A., Van der Wegen, M., 2008. Effect of sea level rise on inlet morphology, COPEDEC VII, Dubai, UAE.

Dronkers, J., 1986. Tidal Asymmetry and estuary morphology, *Netherlands Journal of Sea Research* 20 (2/3), 117-131.

Dronkers, J., 1998. Morphodynamics of the Dutch Delta. Physics of Estuaries and Coastal Seas, Dronkers & Scheffers (eds), Balkema, Rotterdam.

Dronkers, J., 2005. Dynamics of Coastal Systems, Advanced Series on Ocean Engineering –Volume 25, 235-240.

Duc, A.H., 2008. Salt intrusion, tides and mixing in multi-channel estuaries, PhD Thesis, UNESCO-IHE Institute for Water Education, Delft, The Netherlands.

Ehlers, J., 1988. The morphodynamics of the Wadden Sea, Balkema, Rotterdam.

Elias, E. 2006. Morphodynamics of Texel Inlet, IOS Pres, PhD Thesis, Technical University of Delft, ISBN 1-58603-676-9.

Elias, E. P. L., Stive, M. J. F., Bonekamp, J. G. and Cleveringa, J. 2003. Tidal inlet dynamics in response to human intervention, *Coastal Engineering* 45(4), 629-658.

Engelund, F., Hansen, E., 1967. A monograph on sediment transport in alluvial streams, Teknisk Forlag, Copenhagen.

Escaravage, V., Blok, D., Dekker, A., Engelberts, A., Hartog, E., Van Hoesel, O., Schaars, L.K., Markusse, R., Sistermans, W., 2009. Prof Zandsuppletie Oosterschelde het Macrobenthos van de Galgeplaat in het najaar van 2009, Monitor Taakgroep van het NIOO/CEME Rijkswaterstaat Directie Zeeland (in Dutch).

Escofier, F.F., 1940. The Stability of tidal inlets. Shore and Beach. Vol. 8, No. 4, 114-115.

Eysink, W.D. 1993. ISOS*2 Project : Impact of sea level rise on the morphology of the Wadden Sea in the scope of its ecological function, phase 4, General conditions on Hydraulic conditions, Sediment transport, Sand balance, Bed composition and Impact of sea level rise on tidal flats, Delft Hydraulic report H1300, Delft, The Netherlands.

Eysink, W.D. 1991. ISOS*2 Project : Impact of sea level rise on the morphology of the Wadden Sea in the scope of its ecological function, phase 1, Delft Hydraulic report H1300, Delft, The Netherlands.

Eysink, W.D., 1990. Morphologic response of tidal basins to changes, *Coastal Engineering*, 1948-1961.

Eysink, W.D. and Biegel, E.J., 1992. ISOS*2 Project, Impact of sea-level rise on the morphology of the Wadden Sea in the scope of its ecological function, Investigation on empirical morphological relations, Delft Hydraulics, Report H1300.

Fenster, M. and Dolan, R., 1996. Assessing the Impact of Tidal Inlets on Adjacent BarrierIslandShorelines, *Journal of Coastal Research* 12 (1), 294-310.

FitzGerald, D.M., 1988. Shoreline erosional-depositional processes associated with tidal inlets, Lecture Notes on Coastal and Estuarine Studies, Vol, 29, D.G. Aubrey, L. Weishar (Eds.), Hydrodynamic and Sediment Dynamics of tidal inlets, Spring-Verlag New York, Inc., 186-225.

FitzGerald, D.M., 1977. Hydraulics, Morphology and Sediment Transport at Price Inlet, South Carolina, Ph.D. dissertation, Geology Dept., Uni. South Carolina, 84.

FitzGerald, D.M., Hubbard, D.K., Nummedal, D., 1978. Shoreline changes associated with tidal inlets along the South Carolina Coast, In. ASCE (Ed), Proc. 15th Coastal Engineering Conference, 1973-1994.

Fitzgerald, D. M., Kraus, N. C. and Hands, E. B. 2000. Natural mechanisms of sediment bypassing at tidal Inlets, Report ERDC/CHL CHETN-IV-30, US Army Corps of Engineers.

FitzGerald, D.M., Nummedal, D. and Kana, T.W., 1976. Sand circulation pattern at Price Inlet,South Carolina. In: Proc. 15th Coastal Eng. Conf., Honalulu, ASCE, New York, Vol. II, 1868-1880.

FitzGerald, D.M., Penland, S., Nummedal, D., 1984. Control of barrier island shape by inlet sediment bypassing: East Frisian Islands, West Germany, Journal of Marine Geology 60, 355-376.

Friedrichs, C.T., Aubrey, D.G., 1988. Non-linear Tidal distortion in Shallow Well-mixed Estuaries: a Synthesis, *Journal of Estuarine, Coastal and Shelf Science* 27, 521-545.

Friedrichs, C.T., Aubrey, D.G., Speer, P.E., 1990. Impacts of Relative Sea-level Rise on Evolution of Shallow Estuaries, Coastal and Estuarine Studies, Vol. 38, R.T. Cheng (Ed.), Residual Currents and Long-term Transport, Springer-Verlag, New York.

Fry, V.A., Aubrey, D.G., 1990. Tidal Velocity Asymmetries and Bedload transport in Shallow Embayments, *Journal of Estuarine, Coastal and Shelf Science* 30, 453-473.

Galappatti, R., 1983. A depth integrated model for suspended transport, Report 83-7, Communications on Hydraulics, Department of Civil engineering, Delft University of Technology.

Galgano, F.A., 2009b. Beach erosion adjacent to stabilized microtidal inlets, *Middle States Geographer*, 42, 18-32.

Galgano, F.A., 2009a. Shoreline Behavior Along the Atlantic Coast of Delaware, *Middle States Geographer*, 41: 74-81.

Gao, S., and Jia, J. J., 2003. Sediment and carbon accumulation in a small tidal basin: Yuchu, Shandong Peninsula, China, *J. Regional Environmental change*, Vol. 4, 1,63-69.

Gerritsen, H., Berentsen, C.W.J., 1998. A modelling study of tidally induced equilibrium sand balances in the North Sea during the Holocene, *Journal of Continental Shelf Research* 18, 151-200.

Gerritsen, F., De Jong, H., and Langerak, A., 1990. Cross-sectional stability of estuary channels in the Netherlands, Proc. 22nd Int. Conf. on Coastal Engineering, ASCE, 2922-2935.

Gibeaut, J.C., Davis, R.A., 1993. Statistical Geomorphic Classification of Ebb-Tidal Deltas Along the West-Central Florida Coast, *Journal of Coastal Research*, SI 18, 165-184.

Glaeser, D.J., 1978. Global distribution of barrier islands in terms of tectonic setting, *Journal of Geology* 86, 283 – 297.

Groen, P., 1967. On the residual transport of suspended matter by an alternating tidal current,Netherlands *Journal of Sea Research* 3, 4, 564-574.

Groeneweg, J., Van der Westhuysen, A., Van Vledder, G., Lansen, J., Van Dongeren, A., 2008. Wave Modelling in a tidal inlet: Performance of SWAN in the Wadden Sea, Proc. 31st International Conference on Coastal Engineering, 411-423.

Grunnet, N.M., Ruessink, B.G., Walstra, D.J.R., 2005. The influence of tides, wind and waves on the redistribution of nourished sediment at Terschelling, The Netherlands, *Coastal Engineering* 52, 617-631.

Grunnet, N.M., Walstra, D.J.R., Ruessink, B.G., 2004. Process-based modelling of a shoreface nourishment, *Coastal Engineering* 51, 581-607.

Hamm, L., Capobianco, M., Dette, H.H., Lechuga, A., Spanhoff, R., Stive, M.J.F., 2002. A summary of European experience with shore nourishment, *Coastal Engineering* 47, 237-264.

Hanson, H., Brampton, A., Capobianco, M., Dette, H.H., Hamm, L., Laustrup, C., Lechuga, A., Spanhoff, R., 2002. Beach nourishment projects, practices and objectives – a European overview, *Coastal Engineering* 47, 81-111.

Hayes, M.O., 1979. Barrier island morphology as a function of tidal and wave regime, in proceeding of the coastal symposium of barrier islands, edited by S.P. Leatherman, Academic press New York, 1-28.

Hayes, M.O., Goldsmith, V., Hobb, C.H., 1970. Offset coastal inlets, Proc. 12th Coastal Engineering Conference, 1197-1200.

Hibma, A., Stive, M.J.F., Wang, Z.B., 2004. Estuarine morphodynamic, *Coastal Engineering* 51, 765-778.

Hicks, D.M. and Hume, T.M., 1996. Morphology and size of Ebb Tidal deltas at Natural Inlets on pen-sea and Pocket-bay Coasts, North Island, New Zealand, *Journal of Coastal Research* 12, 47-63.

Hicks, M.D., Hume, T.M., Swales, A., and Green, M.O., 1999. Magnitude, spatial extent, time scales and causes of shoreline change adjacent to ebb-tidal delta, Katikati Inlet, New Zealand, *Journal of Coastal Research* 15, 220- 240.

Holgate, S.J., 2007. On the decadal rate of sea level change during the twentieth century, *Geophysical Research Letter*, 34, L01602.

Holthuijsen, L. H., Zijlema, M., Van der Ham, P.J., 2008. Wave Physics in a tidal inlet, Proc. 31st International Conference on Coastal Engineering, 437-447.

Houghton, G.T., Ding, Y., Griggs, D.J., Noguer, M., Van der Linden, P.J., Dai, X., Maskell, K., Johnson, C.A., 2001. Climate Change Scientific Basis, Contribution of Working Group I to the third Assessment report of the Intergovernmental Panel of Climate Change (IPCC), Cambridge University Press, UK, pp. 74-77

Hubbard, D. K., 1976. Changes in inlet offset due to stabilization, In: Proc. 15th Coastal Eng.Conf., Honalulu, ASCE, New York, Vol. II, 1812-1823.

Ikeda, S., 1982. Lateral Bed-Load Transport on Side Slopes. Journal Hydraulics Division, ASCE, Vol. 108, No. 11.

Israel, C.G., 1998. Morphologische ontwikkeling Amelander Zeegat, Werkdocumet RIKZ/OS-98.147X.

Jarret, J.T., 1976. Tidal prism-inlet relationships. Gen. Invest. Tidal inlets Rep. 3, 32 pp, US Army Coastal Engineering and Research Centre. Fort Belvoir, Va.

Jelgersma, S., 1979. Sea-level changes in the North Sea basin. In: Oele, E., Schuttenhelm, R.T.E., Wiggers, A.J., (Eds), The Quaternary History of the North Sea, Acta Universitatis Upsaliensis, Uppsala, Sweden, 233-248.

Kana, T.W., Hayter, E.J. and Work, P. A., 1999. Mesoscale Sediment Transport at Southeastern   U.S.Tidal Inlets: Conceptual Model Applicable to Mixed Energy Settings, *Journal of Coastal Research* 15(2), 303-313.

Kiden, P., 1995. Holocene sea level change and crustal movement in the southwestern Netherlands, *Marine Geology*, 124, 21-41.

Kraus, N.C., 2000. Reservoir model of ebb-tidal shoal evolution and sand bypassing, *Journal of Waterway, Port, Coastal and Ocean Engineering*, 126(3), 305-313.

Kraus, N.C., 1998. Inlet cross-sectional area calculated by process-based model, *Coastal Engineering*, 3265-3278.

Kragtwijk, N.G., Zitman, T.J., Stive, M.J.F., Wang, Z.B., 2004. Morphological response of tidal basins to human interventions, *Coastal Engineering* 51, 207-221.

Latteux, B., 1995. Techniques for long-term morphological simulation under tidal action, *Marine Geology* 126, 129-141.

Leendertse, J.J., R.C. Alexander and S-K. Liu, 1973. A three-dimensional model for estuaries and coastal seas,The Rand Corporation (Santa Monica):Volume I : Principles of Computations, R-1417-OWRT, Volume II : Aspects of computation, R-1764-OWRT, Volume III : The interim program, R-1884-OWRT,Volume IV : Turbulent energy computation, R-2187-OWRT.

Lesser, G., 2009. An approach to medium-term coastal morphological modelling, PhD Thesis, UNESCO-IHE Institute for Water Education, Delft, The Netherlands.

Lesser, G., Roelvink, J.A., Van Kester, J.A.T.M., Stelling, G.S., 2004. Development and validation of a three-dimensional morphological model, *Coastal Engineering* 51, 883-915.

Lorenz, E. N., 1972. Predictability: Does the flap of a butterfly's wings in Brazil set off a tornado in Texas?, American Asso. for the Advancement of Science Annual Meeting Prog., 139.

Louters, T., Gerritsen, F., 1994. The Riddle of sands; A Tidal System's Answer to a Rising Sea Level. Report 94.040. RIKZ, The Hague

Louters, T., Gerritsen, F., 1995. The Riddle of sands; Technical report, National Institute for Coastal and Marine Management, Rijkswaterstaat, The Hague, ISBN90-369-0084-0.

Ludwig, G., Müller, H., Streif, H., 1981. New dates on Holocene sea level changes in the German Bight, Spec. Pull. Int. Association Sedimentol, 5, 211-219.

Mann, D., 1993. A Note on Littoral Drift Budgets and Sand Management at Inlets. *Journal of Coastal Research,* 18: 301-308.

Marciano, R., Wang, Z.B., Hibma, A., De Vriend, H.J., Defina, A., 2005. Modelling of channel patterns in short tidal basins, *Journal of Geophysical Reserach*, Vol.100, F01001, doi:10.1029/2003JF000092.

Marquenie, J.M., De Vlas, J., 2005. The impact of subsidence and sea level rise in the Wadden Sea: Prediction and field verification, Springer Berlin Heidelberg, 355-363

Michel, D and Howa, H.L., 1997. Morphodynamic behaviour of a tidal inlet system in a   mixed-energy environment, Phys. Chem. Earth, Vol. 22, No. 3-4, 339-343.

Müller, J.M., Zitman, T., Stive, M., Niemeyer, H.D., 2007. Long-Term Morphological Evolution of the Tidal Inlet, Norderneyer Seegat. Proc. 30th Int. Conf. on Coastal Engineering, ASCE, 4035-4045.

O'Brien, M.P., 1969. Equilibrium flow areas of inlets on sandy coasts, Journal of the Waterways and Harbors division, Proc. ASCE, 43-52.

O'Brien, M.P., 1931. Estuary Tidal Prisms Related to Entrance Areas, Civil Engineering, Vol1, No.8,738-739.

Oertel, G.F., Kraft, J.C., Kearney, M.S., Woo, H.J., 1992. A rational theory for barrier-lagoon development. In. Quaternary Coasts of the United States: Marine and Lacustrine Systems. SEPM Spec. Publ., 48: 77-87.

Oertel, G. F. 1988. Processes of sediment exchange between tidal inlets, ebb deltas and barrier islands, lecture notes on coastal and estuarine studies, Vol. 29; Hydrodynamics and sediment dynamics of tidal inlets, L.W.D.G. Aubrey and L. Weishar, Eds., Springer-Verlag, New York.

Ortiz, C.A.C., 1994. Sea level rise and its impact on Bangladesh, Journal of Ocean and Coastal Management, 23, 249-270.

Pedrozo-Acuna, A., Simmonds, D.J., Otta, A.K., Chadwick, A.J., 2006. On the cross-shore profile change of gravel beaches, *Coastal Engineering* 53, 335-347.

Pingree, R.D., Griffiths, D.K., 1979. Sand transport paths around the British Isles resulting from M2 and M4 tidal interactions. Journal of Marine Biology Associaltion, U.K., 59, 467-513.

Pugh, D.T., 1987. Tides, Surges and mean sea level rise: a hand book for engineers and scientists (472). Chichester: Wiley.

Rahmstorf, S., Cazanave, A., Church, J., Hansen, J., Keeling, R., Parker, D., Somerville, R., 2007. Recent Climate Observations Compared to Projections, *Science*, 316, 709.

Rakhorst, H.D., 2000. Erosie en Sementatie in de buitendelta van het Zeegat van het Vlie en de aangrenzende kuststroken 1926-1984, Deerlondezoek nr. 11, Supplement deerlondezoek nr. 7, Rapportage ANV-2000-20, RWS, directie, Noord-Holland.

Ranasinghe, R., Swinkels, C., Luijendijk, A., Roelvink, J.A., Li, L., Bosboom, J., Stive, M.J.F. and Walstra, D.J., 2010. Morphodynamic upscaling with the MORFAC approach, Books of abstract, 32nd Int. Conf. Coastal Engineering, China, Paper No. 325.

Renger, E. and Partenscky, H. W. 1974. Stability criteria for tidal basins, proc. 14th Coastal Engineering Conf., ASCE, Vol. 2, 1605-1618.

Ridderinkhof, H., 1988. Tidal and residual flows in the Western Dutch Wadden Sea 1: Numerical model results, Netherlands Journal of Sea Research 22(1), 1-21.

Rijzewijk, L.C., 1981. Overzichtskaarten Zeegaten van de Waddenzea 1796-1985, Verzameling 86.H208.

Rinaldo, A., Lanzoni, S. and Marani, M. 2001. River and tidal networks, In: Seminara, G., Blondeaux, P. (Eds.), River, Coastal and Estuarine Morphodynamics. Springer-Verlag, Berlin.

Roelvink, J.A., 2006. Coastal morphodynamic evolution techniques, *Coastal Engineering* 53, 277-287.

Roelvink, J.A. and Reniers, A.J.H.M., 2011. A guide to modelling coastal morphology, Advances in Coastal and Ocean Engineering, World Scientific.

Roelvink, J.A., Reniers, A., Van Dongeren, A., De Vries, J.V.T., Lescinski, J. and Walstra, D.J., 2007. Proc. 10th Workshop on Waves and Coastal Hazards, North Shore, Oahu.

Roelvink, D., Reniers, A., Van Dongeren, A., De Vries, J.V.T., McCall, R., Lescinski, J., 2009. Modelling storm impacts on beaches, dunes and barrier islands, *Coastal Engineering* 56, 1133-1156.

Roelvink, J.A., Walstra, D.J., 2004. Keeping it simple by using complex models, Advances in Hydroscience and Engineering, Volume V1, 1-11.

Roelvink, J.A., Van der Kaaij, T., Ruessink, B.G., 2001. Calibration and verification of large-scale 2D/3D flow models, Phase 1, Delft Hydraulics report, Z3029.11.

Ruessink, B.G., Walstra, D.J.R., Southgate, H.N., 2003. Calibration and verification of a parametric wave model on barred beaches, *Coastal Engineering* 48, 139-149.

Ruggiero, P., Walstra, D.J.R., Gelfenbaum, G., Van Ormondt, M., 2009. Seasonal-scale nearshore morphological evolution: Field observations and numerical modelling, *Coastal Engineering* 56, 1153-1172.

Sha, L.P., 1989. Variation in ebb-delta morphologies along the West and East Frisian Islands, the Netherlands and Germany, *Marine Geology* 89, 11-28.

Sha, L.P., 1989b. Sand transport patterns in the ebb-tidal delta off Texel Inlet, Waddsen Sea, The Netherlands, *Marine Geology* 86, 137-154.

Sha, L.P., Van den Berg, J.H., 1993. Variation in ebb-tidal delta geometry along the coast of The Netherlands and the German Bight, *Journal of Coastal Research* 9(3), 730-746.

Soulsby, R.L., 1997. Dynamics of marine sands: A manual for practical applications, Thomas Telford, HR Wallingford, London.

Speer, P. E., Aubrey, D.G. and Friedrichs, C. 1991. "Nonlinear hydrodynamics of shallow tidal inlet / bay systems" Tidal Hydrodynamics, B. B. Parker, ed., John Wiley, New York, 321-339.

Speer, P.E., Aubrey, D.G., 1985. A Study of non-linear tidal propagation in shallow Inlet/Estuarine Systems, Part II: Theory, *Journal of Estuarine, Coastal and Shelf Science* 21, 207-224.

Steetzel, H., 1995. Voorspelling ontwikkeling kustlijn en buiten-delta's waddenkust over de periode 1990-2040. Delft Hydraulics report no. H1887 prepared for Rijkswaterstaat, The Hague, The Netherlands (in Dutch).

Stelling, G.S., 1984. On the construction of computational methods for shallow water flow problems. Rijkswaterstaat communications, No. 35, The Hague, Rijkswaterstaat, 1984.

Stelling, G.S. and J.J. Lendertse, 1991. Approximation of Convective Processes by Cyclic ACI methods, Proceeding 2nd ASCE Conference on Estuarine and Coastal Modelling, Tampa, 1991.

Stive, M.J.F., Capobianco, M., Wang, Z.B., Ruol, P., Buijsman, M.C., 1998. Morphodynamics of a Tidal Lagoon and the Adjacent Coast, 8th International Biennial Conference on Physics of Estuaries and Coastal Seas, The Hague, September 1996, 397-407.

Stive, M.J.F., Nicholls, R.J. and De Vriend, H.J., 1991. Sea-level rise and shore nourishment: a discussion, *Coastal Engineering* 16, 147-163.

Stive, M.J.F., Roelvink, J.A., and De Vriend, H.J., 1990. Large-scale coastal evolution concept, Proc. 22nd Int. Conf. on Coastal Engineering, ASCE, 1962-1967.

Sutherland, J., Peet, A.H., and Soulsby, R.L., 2004. Evaluating the performance of morphological models, *Costal Engineering* 51, 917-939.

Thijsse, J.T., 1972. Een halve Zuider Zeewerken 1920-1970, Tjeenk Willink, Groningen.

Van der Kreeke, J., Robaczewska, K., 1993. Tide-induced residual transport of coarse sediment; application to the Ems estuary, *Netherlands Journal of Sea Research* 31, 209-220.

Van der Meij, J.L., Minnema, B., 1999. Modelling on the effect of a sea-level rise and land subsidence on the evolution of the ground water density in the subsoil of the northern part of the Netherlands, *Journal of Hydrology* 226, 152-166.

Van der Molen, J., and De Swart, H.E., 2001. Holocene tidal conditions and tide-induced sand transport in the Southern North Sea, *Journal of Geophysical Research*, 106, 9339-9362.

Van der Molen, J., and Van Dijk, B., 2000. The evolution of the Dutch and Belgian coasts and the role of sediment supply from the North Sea, *Journal of Global and Planetary Change*, 27, 223-244.

Van der Speck, A.J.F., 1994. Large-scale evolution of Holocene tidal basins in the Netherlands, PhD Thesis, University of Utrecht, ISBN 90-393-0664-8.

Van der Wegen, M., Thanh, D.Q., Roelvink, J.A., 2006. Bank erosion and morphodynamic evolution in alluvial estuariesnusing a process based 2D model, 7th Conf. Hydroscience and Engineering, Philadelphia, USA.

Van der Wegen, M., Roelvink, J.A., De Ronde, J., Van der Spek, A., 2008. Long-term morphodynamic evolution of the Western Scheldt estuary, The Nethelands, using a process-based model, COPEDEC VII, Dubai, UAE.

Van der Wegen, M., Roelvink J. A., 2008. Long-term morphodynamic evolution of a tidal embayment using a two-dimensional, process-based model, *Journal of Geophysical Research*, 113, C03016, doi:10.1029/2006JC003983.

Van Dongeren, A.R., De Vriend, H.J., 1994. A model of morphological behaviour of tidal basins, *Coastal Engineering* 22, 287-310.

Van Duin, M.J.P., Wiersma, N.R., Walstra, D.J.R., Van Rijn, L.C., Stive, M.J.F., 2004. Nourishing the shoreface: observation and hindcasting of the Egmond case, The Netherlands, *Coastal Engineering* 51, 813-837.

Van Geer, P.F.C., 2007. Long-term morphological evolution of the Western Dutch Wadden Sea, M.Sc. Thesis, Technical University of Delft, The Netherlands.

Van Goor, M.A., 2001. Influence of Relative Sea Level Rise on Coastal Inlets and Tidal Basins, M.Sc. Thesis, Technical University of Delft, The Netherlands.

Van Goor, M.A., Zitman, T.J., Wang, Z.B., Stive, M.J.F., 2003. Impact of sea-level rise on the morphological equilibrium state, *Marine Geology* 202, 211-227.

Van Leeuwen, S.M., Van der Vegt, M., De Swart, H.E., 2003. Morphodynamics of ebb-tidal deltas: a model approach, *Journal of Estuarine, Coastal and Shelf Science* 57, 1-9.

Van Rijn, L.C., 1993. Principles of sediment transport in rivers, estuaries and coastal seas. AQUA Publications, the Netherlands.

Van Rijn, L. C., Walstra, D.J.R., Grasmeijer, B., Sutherland, J., Pan, S., Sierra, J.P., 2003. The predictability of cross-shore evolution of sandy beaches at the scale of storm and seasons using process-based profile models, *Coastal Engineering* 47, 295-327.

Van Veen, J., 1936. Onderzoekingen in de hoofden, in verband met de gesteldheid der Nederlandsche kust (in Dutch), Algemeene Landsdrukkerij's Gravenhage, the Netherlands.

Van Vledder, G., Groeneweg, J., Van der Westhuysen, A., 2008. Numerical and Physical aspects of wave modelling in a tidal inlet, Proc. 31st International Conference on Coastal Engineering, 424-435.

Walton, T.L., Adams, W.D., 1976. Capacity of inlet outer bars to store sand, In: Proc. 15th Coastal Eng. Conf., Honolulu, ASCE, New York, Vol. II, 1919-1937.

Wang, Z.B., Jeuken, C., De Vriend, H.J., 1999. Tidal asymmetry and residual sediment transport in estuaries. A literature study and applications to the Western Scheldt, WL|Delft Hydraulics report Z2749, Delft, the Netherlands.

Wang, Z. B., Karssen, B., Fokkink, R. J. and Langerak, A., 1998, A dynamic-empirical model for estuarine morphology, in J. Dronkers and M.B.A.M. Scheffers (eds.), Physics of estuarine and coastal seas. Balkema, Rotterdam, 279-286.

Wang, Z.B., Louters, T. and De Vriend, H.J., 1995. Morphodynamic modelling for a tidal inlet in the Wadden Sea, *Marine Geology* 126, 289-300.

Wilkens, J., 1998. Bar morphology Bornrif, Delft Hydraulics Report, Z2526/K2000*BR.

Work, P.A., and Dean, R.G., 1990. Even/Odd analysis of shoreline changes adjacent to Florida's tidal inlets. Proc. 22nd International Coastal Engineering Conference, 3, 2522-2535.

Zagwijn, W.H., 1986. Netherlands in het Holoceen, Rijks Geologische, Dienst Haarlem, Staatsuitgeverij, s-Gravenhage.

Zhang, K., Douglas, B.C., Leatherman, S.P., 2004. Global warming and coastal erosion, *Journal of Climate Change*, 64, 41-58.

List of Figures

List of Tables

Publications

Peer-reviewed publications

1. Dissanayake, D.M.P.K., Van der Wegen, M., and Roelvink, J. A, 2009c. Modelled channel pattern in schematized tidal inlet, *Coastal Engineering* , 56, 1069 – 1083.

2. Dissanayake, D.M.P.K., Ranasinghe, R., Roelvink, J.A., (*in press*). The morphological response of large tidal inlet/basin systems to Relative Sea level rise, *Climatic Change*.

3. Dissanayake, D.M.P.K., Ranasinghe, R., Roelvink, J.A., Wang, Z.B., (*under review*). Process based vs Scale aggregated modelling of long term evolution of large tidal inlet/basin systems, *Coastal Engineering*.

Conference attendance and proceedings

1. Dissanayake, D.M.P.K., Ranasinghe, R., Roelvink, J.A., (*accepted*). Numerical approach on the inlet effects on adjacent coastlines, 8th International Conference on Coastal and Port Engineering in Developing Countries, Chennai, India.

2. Dissanayake, D.M.P.K., Ranasinghe, R., Roelvink, J.A. and Wang, Z.B., 2011. Process-based and Semi-empirical modelling approaches on tidal inlet evolution, *Journal of Coastal Research*, SI 64, Proc. of the 11th International Coastal Symposium, Szczecin, Poland, pp.1013-1071.

3. Roelvink, D., Van der Wegen, M., Dastgheib, A., Dissanayake, P. and Ranasinghe, R., 2010. Predictability of the main channel and shoal pattern of estuaries and tidal inlets, Proc. 15th Physics of Estuaries and Coastal Seas, Colombo, Sri Lanka, *CD Rom.*

4. Van der Wegen, M., Roelvink, D., Ranasinghe, R., Dastgheib, A. and Dissanayake, P., 2010. Sea level rise and morphodynamic evolution of tidal basins. Proc Int Conference Deltas in Times of Climate Change. Rotterdam, the Netherlands.

5. Dissanayake, D.M.P.K., Ranasinghe. R., and Roelvink, J.A., 2009b. Effect of sea level rise in inlet evolution: a numerical modelling approach, *Journal of Coastal Research*, SI 56, Proc. of the 10th International Coastal Symposium, Lisbon, Portugal, pp. 942- 946.

6. Dissanayake, D.M.P.K., Roelvink, J.A., Ranasinghe, R., 2009. Influence of tidal inlet along the adjacent coastlines, NCK Days, Texel, the Netherlands, *poster presentation.*

7. Dissanayake, D.M.P.K., Roelvink, J.A., Ranasinghe, R., 2009a. Process-based approach on tidal inlet evolution – Part II, Proc. International Conference in Ocean Engineering, Chennai, India. *CD Rom.*

8. Dissanayake, D.M.P.K., Roelvink, J.A., Ranasinghe, R., 2008. Influence of sea level rise on inlet morphology, Dynamics Days in Europe, Delft, the Netherlands, *poster presentation.*

9. Dastgheib, A., Van der Wegen, M., Dissanayake, D.M.P.K., Roelvink, J.A., 2008. Long-term process-based morphological modelling of tidal basins and estuaries in the Netherlands, Proc. of the 31st International Conference on Coastal Engineering, Hamburg, Germany, Vol. III, pp. 2231-2243.

10. Dissanayake, D.M.P.K., Roelvink, J.A. and Van der Wegen, M., 2008. Effect of sea level rise on inlet morphology, 7th International Conference on Coastal and Port Engineering in Developing Countries Dubai, United Arab Emirates, *CD Rom*

11. Dissanayake, D.M.P.K. and Roelvink, J.A., 2008. Effect of sediment transport formulation on long-term morphology, European Geophysical Union (EGU), Vienna, Austria.

12. Dissanayake, D.M.P.K. and Roelvink, J.A., 2008. Effect of tidal asymmetry and sea level rise on inlet morphology, NCK Days, Delft, the Netherlands.

13. Dissanayake, D.M.P.K. and Roelvink, J.A., 2007. Process-based approach on tidal inlet evolution – Part 1, Proc. 5th IAHR Symposium on River, Coastal and Estuarine Morphodynamics, Twente, the Netherlands, pp. 3-9.

14. Dissanayake, D.M.P.K. and Roelvink, J.A., 2007. Effect of Sediment transport formulation on long-term morphology, NCK Days, Kijkduin, the Netherlands, *poster presentation.*

About the Author

Pushpa Kumara Dissanayake was born on the 15th of July 1974 in Onegama, Sri Lanka. He obtained his primary education in Hingurakgoda. In 1990, he moved to Kandy and completed his secondary education. He studied civil engineering at the University of Peradeniya from 1995 until 1999 graduating with second class honours. From 2000 to 2003, he worked as a research engineer for Lanka Hydraulic Institute and mainly engaged in numerical modelling of coastal projects. For optimizing the layout of Colombo Port South Harbour Development Project, for hydrodynamics and wave transformation of Coastal Resources Management Project and for preventing sandbar formation of Madu Ganga Outfall are some of them.

He travelled to the Netherlands in 2003 after obtaining a post-graduate scholarship to study at the UNESCO-IHE Institute for Water Education, which was funded by the Netherlands Fellowship Programme. He studied in Coastal Engineering and obtained MSc in 2005. His thesis project, undertaken at WL|Delft Hydraulics (presently Deltares), investigated the effect of bed roughness formulations on morphodynamics of the Western Scheldt Estuary, the Netherlands. After graduation, he returned to Sri Lanka and worked for Lanka Hydraulic Institute as a senior modelling engineer, where he was responsible for some overseas projects including Wave influence on dredged material containment area at Busan harbour (Korea), Wave transformation at Palm Tree Islands (Maldives), 2D physical modelling on Al Marjan Island (UAE) and environmental impact assessment of the proposed Sethu Samudram Canal Project between Sri Lanka and India.

In February 2006, he started his PhD research at UNESCO-IHE on the responding morphodynamic evolution of large tidal inlet/basin systems to sea level rise, which was funded by Delft Cluster Research Programme and resulted in the present thesis. In January 2010, he travelled to Norderney, Germany where he now lives with his wife Madhuka and two daughters Thamodie (4) and Mahelie (7 years). On Norderney, he completed the final preparation of this thesis while working with his present employer Forschungsstelle Küste - NLWKN where he works on climate change adaptation strategies on the Lower Saxony Coast, East Frisian Wadden Sea.

T - #0149 - 160425 - C196 - 244/170/11 - PB - 9780415621007 - Gloss Lamination